COPPER DEPOSITS:

Origin, Evolution, and Present Characteristics

An analysis of the present copper deposits in terms of the geometrical, mechanical, thermal, and chemical aspects of the Earth's behavior during the past 4.6 billion years.

J. H. TATSCH

TATSCH ASSOCIATES

Sudbury, Massachusetts 01776

1975

Printed in the United States of America, June, 1975. Copies are available from Tatsch Associates or from book dealers in all parts of the world.

International Standard Book Number: 0-912890-08-8.

Library of Congress Catalog Card Number: 74-78916.

Preface

Because most copper deposits that display easily recognizable surficial manifestations have already been found, the future success in copper exploration will depend increasingly upon a more sophisticated approach, preferably one that embodies a clear understanding of the genetic relationships that exist between mineral deposits and geodynamic processes and phenomena. One such approach is the subject of this book: it is based on the observation that economic concentrations of minerals are not randomly distributed but appear to be related to tectonic processes and regimes within the Earth. The details of these genetic relations are not known, and speculations regarding them have long been the subject of discussion among geologists. The resulting diversity of opinion can be seen by reading any applicable journal or by attending any meeting devoted to economic geology.

Copper Deposits, briefly stated, summarizes the results of an independent research project undertaken to determine the most probable geometrical, mechanical, thermal, and chemical aspects of the origin, evolution, and present characteristics of the Earth's copper deposits. This book shows how these deposits have originated, have evolved, and have become emplaced into their present distribution patterns in accordance with a single, long-lived, deep-seated, global mechanism that has been operating within the Earth during the 4.6 billion years that the Earth is believed to have been in existence. This mechanism is the "tectonospheric Earth model", which has been described in the book, *The Earth's Tectonosphere,* and used as the basis for the books, *Mineral Deposits* and *Petroleum Deposits,* by the same author.

One of the strengths of the Tectonospheric Earth Model is that it permits the prediction of certain variations that have occurred within the geometrical, mechanical, thermal, and chemical heterogeneities of the Earth's upper 1000 km during the past 4.6 billion years. The ability of the model to provide these predictions forms the basis for its use as a supplementary tool in the exploration for copper deposits.

Table of Contents

Chapter 1

COPPER DEPOSITS: A GLOBAL SURVEY

The Earth's mineral deposits form certain unique patterns within the spatio-temporal framework through which these deposits have evolved during the 4.6 billion years (b. y.) that the Earth is believed to have been in existence. There are geographical regions in which minerals are found to be concentrated as much as four orders of magnitude above the average crustal abundances. Gold, copper, and zinc, for example, are found concentrated three orders of magnitude above the average crustal abundances for these metals; lead, four orders of magnitude above the average. Some of these mineralization provinces are well known: Zambia (copper), Lake Superior (copper), and South Africa (gold). Other mineralization provinces are becoming better known as the need for scarce minerals continues to increase.

The purpose of this book is the analysis of copper deposits on a global basis in terms of the geometrical, mechanical, thermal, and chemical aspects of the Earth's behavior that appear to have contributed to the origin, evolution, and present characteristics of these deposits. Ambiguous evidence makes it difficult to define the source of minerals (See, e. g., Goldschmidt, 1954; Levorsen, 1967; Krauskopf, 1971). What is known is that the processes involved in the origin, evolution, migration, enrichment, and emplacement of mineral deposits appear to constitute a multistage sequence encompassing thermomagmatic and geodynamic activity, including: (1) subsurficial magmatic intrusion, cooling, and differentiation; (2) surficial volcanism and other extrusive activity;

and (3) a complete spectrum, or continuum, of metamorphic activity, erosion, weathering, and leaching by surficial and subsurficial hydraulic activity.

The point of departure for this analysis is an extension of the concept of the global symmetry of tectonic belts (Tatsch and Talbot, 1971). Because there is no clear-cut, globally-compatible theory for the ultimate source of minerals, it will be assumed in this analysis that the unknown aspects regarding the origin, evolution, migration, and present accumulations of minerals are as predicted by a "tectonospheric Earth model" (Tatsch, 1972a). This model, briefly stated, permits one to make working-hypothesis type predictions about the salient relationships that exist between magma types and geologic structure, between specific mineral types and thermal and stress patterns, between genetically intrusive and extrusive minerals, and between other existing relationships that are not predictable from present models when considering large areas over long periods of time. The Tectonospheric Earth Model will be used in this analysis also to provide the basic frame of reference for making predictions regarding the geometrical, mechanical, thermal, and chemical behavior of the Earth's upper 1000 km during the past 4.6 b. y. in those cases where other applicable concepts are weak, non-committal, or non-existent.

The Global and Cosmic Abundances of Minerals.

The global and cosmic abundances of minerals are not the same. Nickel provides a good example: about eleventh cosmically, about thirteenth within the solar atmosphere, sixth within meteoritic bodies, and fifth within the Earth (after iron, oxygen, silicon, and magnesium).

Similarly, there are large contrasts between global concentrations and crustal concentrations of minerals. Although nickel comprises about 3% of the Earth, the crustal abundances are more than an order of magnitude below this concentration. Some individual, natural iron-nickel alloys have as much as 50% nickel (See, e. g., Boldt, 1967). The richest crustal concentrations of nickel occur in those rocks derived from mantle sources. These are found in nickeliferous provinces distributed roughly as follows:

Canada	57%
U. S. S. R.	21%
New Caledonia	11%
Cuba	4%
United States	3%
South Africa	1%
Others (Albania, Finland, Poland, and Brazil)	3%
	100%

By continent, this distribution is as follows:

North America	64%
Eurasia	23%
Australasia	11%
Africa	1%
South America	1%
	100%

These distributions show that 2/3 of the known surficial nickel is concentrated in one shield-plate area, leaving only 1/3 for the other shield-plate areas of the Earth.

Other minerals show similar anomalies in global distribution patterns. Thus, when the copper deposits of the world are analyzed on a global scale, three primary types of copper deposits evolve, each with characteristic spatio-temporal distribution patterns: (1) porphyry deposits (47.5%); (2) stratabound impregnations in sedimentary rocks (26.5%); and (3) massive pyrite deposits associated with volcanic rocks (10.7%). For these types, the distributions are normal for the grades of copper; log-normal, for the quantities. The correlation between the grades and the logarithms of the volumes are significant: (1) negative for the porphyries; (2) positive for the stratabound impregnations; and (3) approximately zero for the massive pyrites. Closer scrutiny of the data indicates a strong concentration of types within spatio-temporally distinct depositions (Pelissonnier, 1968).

Specific anomalies within copper distribution patterns will be discussed later. Suffice it here to say merely that minerals have not been uniformly deposited either spatially or temporally. One of the purposes of this book is the analysis of certain heterogeneities in the spatio-temporal emplacement of copper deposits during the 4.6 b. y. that the Earth is believed to have been in existence.

Representative Copper Deposits.

It has been known for some time that some regions of the Earth have abundant mineralizations of certain types, such as gold or copper, for example. Thus, Arizona, where copper mineralizations have occurred during many Precambrian and Phanerozoic episodes, is known as a copper province (See, e. g., Titley & Hicks, 1966). The exact cause for the localization of minerals into provinces is not known (See, e. g., Bartholome, 1974; Pelissonnier, 1972). However, a close scrutiny of the spatio-temporal framework within which the contributing mineralizations have occurred suggests that the central mechanism controlling the geometrical, mechanical, thermal, and chemical aspects of the associated seismotectonomagmatic belts has been closely constrained by long-lived deep-seated motivations that have been operating throughout the 4.6 b. y. that the Earth is believed to have been in existence.

Before examining these motivational relationships, it is well to summarize some of the world's representative copper deposits:

a. North America.

The 15,000-km American segment of the present circum-Pacific orogenic belt, which suggests repreated orogenies during both the Phanerozoic and the Precambrian, produces over half of the world's copper. Most of these deposits are associated with Tertiary batholiths and other intrusives. These deposits contain also gold, silver, molybdenum, lead, and zinc, with concentrations of these lesser minerals generally increasing outward.

b. South America.

Part of the 15,000-km American segment of the present circum-Pacific orogenic belt includes rich deposits in Chile and Peru. Associated with Tertiary monzonite intrusives, these deposits are partially-oxidized sulfide replacements containing bornite, chacocite, and chalcopyrite.

c. Eurasia.

Widely scattered copper deposits constitute the Eurasian provinces: (a) Cornwall, England; (b) Huelva, Spain; (c) Ural Mountains; (d) German Kupferschiefer; (e) the outer Japanese zone.

d. Africa.

The 500-km northwesterly trending belt lying in southwestern Africa constitutes one of the world's largest copper provinces.

e. Australasia.

Many of Australia's minerals fall into provinces that are less well defined than are those of other continents. Among the fairly clearly defined is the gold province, which includes lodes, placers, and saddle reefs of Victoria, New South Wales, and Tasmania. But Australia's copper provinces are not very clearly defined. Those that do exist are mainly in Tasmania.

Most authors accept the concept of mineralization provinces. Most agree also that the present characteristics of metallogenic provinces are indisputable (See, e. g., Spurr, 1923; Lindgren, 1933; Bateman, 1950, 1951; and most authors writing since then). However, there is little agreement on exactly why and how these mineral-laden provinces began and evolved.

Thus, we find many unanswered questions regarding the origin and evolution of mineralization provinces. Why should a 15,000-km (roughly 135°) segment of a great circle along the west coast of the Americas be so heavily endowed with copper? Why does the Canadian shield contain a 35° great-circle belt that is highly auriferous? Why does over 80% of the world's Tertiary petroleum lie along segments of three mutually-orthogonal great circles?

Earlier, it was thought that these elongate mineralization provinces owed their existence to erosive processes that, for example, uncovered the auriferous belts of the Canadian, Siberian, and African shields. But this only compounded the problem by presupposing that by some means or other (1) the minerals were zoned and (2) all the gold zones lay within the same horizon.

Later, some of the simple plate-tectonic concepts offered the suggestion that down-going slabs of lithosphere, as they dug their way into 800-km depths of mantle material, became heated and exuded belts of minerals such as, for example, supposedly, (1) the 15,000-km copper belt of the western Americas, and (2) the

3500-km gold belt of Canada. Many of the geometrical, mechanical, thermal, and chemical aspects of the simple plate-tectonic concept for the origin, evolution, and present characteristics of mineralization provinces remain to be defined.

The details of some of the world's mineralization provinces have been analyzed recently by Brown (1973), Chadhuri and Cullers (1973), Clark and Kelly (1973), Dar (1968, 1972), Dodge (1972), Whiteman et al. (1975), Bartholome (1974), Pelissonnier (1972), Ernst (1972, 1973), Hanor (1973), Jones et al. (1973), Kutina and Fabbri (1972), Lowell and Sides (1973), Miller (1973), Orajaka (1973), Wiltse (1973), Yakovlev and Starostin (1972), Erdman (1972), Maximov (1972), North (1971, 1972), Lowell (1973), Corliss (1973), Comer (1973), Weber and Stephenson (1973), Wilson and Laznicka (1972), and by earlier authors cited by these.

<u>The Geochemistry of Copper Deposits</u>.

Copper falls into group II-B of the periodic table. With the atomic number 29 and an atomic weight of 63.546, it consists of the isotopes 63-Cu (69.1%) and 65-Cu (30.9%). Copper deposits consist of native copper, compounds of Cu(+1) and Cu(+2).

Copper, a strong chalcophile (sulfur-seeking element), tends to concentrate in sulfide deposits. In igneous rocks, it occurs primarily as finely divided sulfides, e. g., chalcopyrite and bornite. Within the igneous cycle, copper is concentrated in basaltic and gabbroic rocks, being most highly concentrated in the ferromagnesian such as the pyroxenes and biotite.

During the crystallization of igneous rocks, copper, along with other elements, tends to be expelled from the structures of silicate minerals. If the magma contains appreciable volatiles, copper may be dissolved and transported with the volatiles along "paths of preferential travel" (Tatsch, 1972a). This causes the formation of veinlets and extensive vein-type copper deposits. In some cases, particularly if the volatile content is low, the copper may remain in the intrusion, concentrating as an immiscible copper sulfide fluid, e. g., some gabbroic rocks. Within igneous rocks, the average abundances of copper may be summarized (in ppm): basaltic and gabbroic rocks (90), andesites (35), ultramafic rocks (15), and granitic rocks (15).

In those cases where copper is carried to the ocean, less than 0.1% remains in solution. The remainder is precipitated, primarily with clays, with manganese oxides, or with biota.

Within all sedimentary rocks, some copper is found even if only in traces (See, e. g., Cox et al., 1973). Average abundances may be summarized (in ppm): shales and clays (45), phosphorites (30), soils (20), coals (15), sandstones (10), and limestones (4). These relative abundances reflect the solubility and mobility of copper in the presence of oxidizing solutions. For example, marine clays and other fine-grained rocks tend to concentrate copper more than do coarse-grained rocks and limestones. Within types of rocks, other concentration preferences are also noted. For example, black shales contain more copper than do other shales. This is particularly noticeable in organic-rich black shales, which may contain 200 ppm of copper.

Pelagic sediments contain significantly more copper than do continental rocks. Some deep-sea clays average over 500 ppm of copper. Some areas (such as the crests of oceanic ridges) may contain twice that amount. Others (such as the Atlantis II deep in the Red Sea) contain an order-of-magnitude greater concentration of copper.

The Formation of Copper Deposits.

Copper is introduced into the Earth's crust, from unknown deeper sources, by (1) igneous intrusions and (2) upward migrating fluids. Enrichments of copper deposits may result from the secondary effects of weathering, erosion, and sedimentation. In many cases, copper deposits are remobilized and reconcentrated by subsequent stages of intrusion and by metamorphism. In other cases, ground and connate waters may dissolve, transport, and redeposit copper to form "younger" deposits.

The weathering of some copper deposits, such as chalcopyrite, oxidizes the copper to ionic state, a portion of which may be adsorbed on organic matter, on clay minerals, and on iron-oxide particles. In some reducing environments, copper may be precipitated as sulfides by sulfur-reducing bacteria, or it may be adsorbed by decaying humic material. In some oxidizing environments, such as in manganese nodules and in pelagic muds, copper oxide deposits form. Some copper may be concentrated in residual brines, but the behavior of copper in an evaporite environment is poorly understood (See, e. g., Cox et al., 1973). Some copper deposits may form from connate pore waters within basins of nonmarine deposition, where the precipitation results from reactions with hydrogen sulfide or with organic matter. In other cases, the copper deposits occur as carbonates, sulfates, silicates, and phosphates.

Some evidence suggests that the geochemical cycle for the formation of copper deposits during the Archean and part of the Proterozoic was significantly different because the Earth's modern atmosphere did not develop until about 1.3 b. y. ago. In such an environment, primary copper sulfide minerals were probably stable enough to survive weathering and redeposition. Consequently many Archean and Proterozoic copper deposits have survived to this date (Chapters 9 and 10). In other cases, some of these earlier deposits were exposed to oxygenated waters and redistributed in chemically reactive environments within sedimentary sequences, such as beds rich in organic matter.

The Mineralogy of Copper Deposits.

Copper deposits comprise over 150 different minerals. Chalcopyrite ($CuFeS_2$) is the most abundant copper sulfide, followed by bornite (Cu_5FeS_4) and chalcocite (Cu_2S). Other major copper minerals are: native copper (Cu), covellite (CuS), enargite (Cu_3AsS_4), cuprite (Cu_2O), and malachite ($Cu_2(OH)_2(CO_3)$). Less-important copper minerals include: tetrahedrite, tennantite, famatinite, stannite, atacamite, tenorite, azurite, chrysocolla, brochantite, antlerite, chalcanthite, and kroehnkite.

Most copper deposits fall into one of five types: (1) porphyry, vein, and replacement deposits, which are genetically associated with felsic intrusive rocks; (2) sedimentary rocks; (3) massive sulfide deposits in volcanic rocks; (4) nickel-copper ores formed by magmatic processes in mafic intrusions; and (5) native copper of the Keweenaw type. These types, and other types of lesser importance, have been discussed repeatedly (See, e. g., Cox et al., 1973). Some will be discussed in later chapters. Suffice it here to say merely that roughly 65% of the copper deposits fall into category 1; 25% into category 2; and 5% into category 3. The other roughly 5% of the Earth's copper deposits fall into categories 4 and 5 and into other types of lesser importance.

Chapter 2

THE TECTONOSPHERIC EARTH MODEL: A NEW CONCEPT

One of the theses of this book is that the Tectonospheric Earth Model provides a supplementary tool in the exploration for copper. It does that, according to this concept, by analyzing the geometrical, mechanical, thermal, and chemical aspects of the substructure and related geophenomena that have contributed to the origin, evolution, and present characteristics of copper deposits. This chapter summarizes those aspects of the Tectonospheric Earth Model that are essential to an understanding of the basic concepts embodied in that model. Greater detail regarding the model is contained in another work (Tatsch, 1972a).

The Basic 5400-km Primordial Earth.

According to the dual primeval planet hypothesis (Tatsch, 1972a), the Earth may be assumed to be composed essentially of a 5400-km primordial Earth onto which 5 octants of Earth Prime have been accreted, where, simply stated, the basic 5400-km primordial Earth is differentiated and fractured into octants. Such a body is shown schematically in Fig. 2-1.

The center, C, represents a solid core, with an undefined radius assumed to lie somewhere within the range 0 to several hundred km. The hatched spherical shell, MC, represents a molten core. The 8 octants of the primordial mantle surround the molten core and are marked S. The preferential flow of heat, and possibly of volatiles, outward from the fracture belts, F, causes igneous activity at I.

The primordial Earth has a volume such that, when 5/8 of an

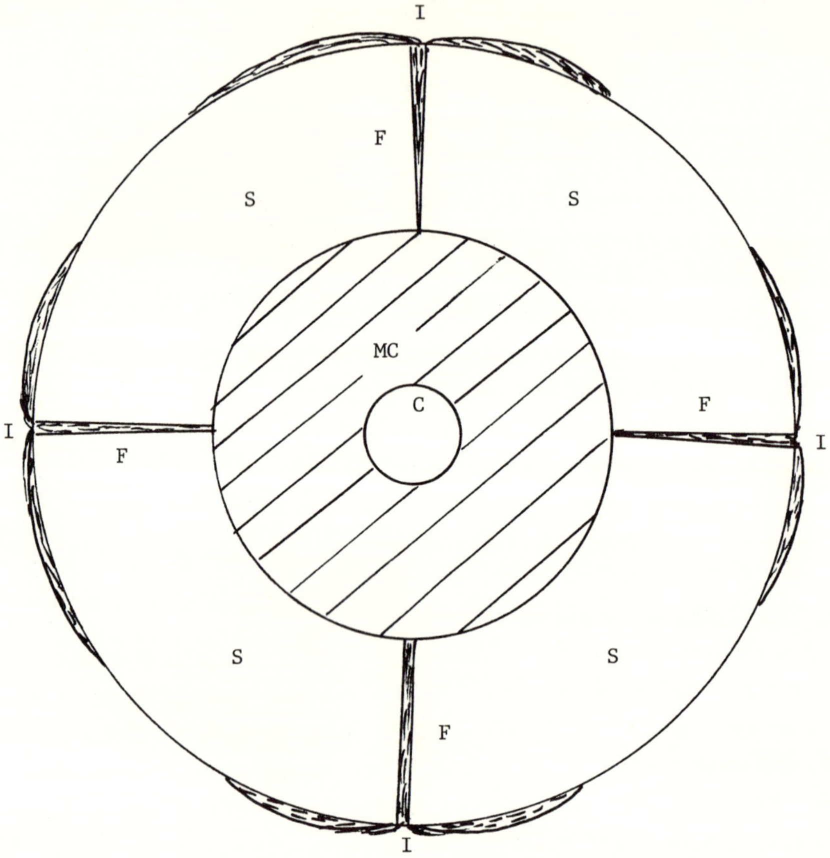

Figure 2-1. The 5400-km primordial Earth which, according to the Tectonospheric Earth Model, has been cut by three mutually orthogonal planes passing through the center. Schematic; not to scale. See text for details.

equivalent volume is added, the resulting body has a radius of about 6371 km, the radius of the actual Earth. This gives the primordial Earth a radius of about 5400 km, thus:

$1\ 5/8\ V' = V$, where V' and V are the volumes of the primordial Earth and of the actual Earth, respectively. Then, since

$$V = 4/3\ \pi R^3, \text{ we may write}$$

$$1.625\ (R')^3 = R^3, \text{ from which}$$

$$R' = (1.625)^{-1/3} R$$

$$R' = (1.625)^{-1/3}\ (6371 \text{ km}) \doteq 5400 \text{ km}.$$

This is the radius of the primordial Earth, as well as of Earth Prime and of the subtectonosphere used in the model.

The thickness of the tectonosphere is about 6371 km minus 5400, equals 971, or about 1000 km, roughly. Although the values, 5400 km and 1000 km, are used in this model for the radius of the subtectonosphere and for the thickness of the tectonosphere, respectively, it should be pointed out that these values are approximations which depend somewhat upon the particular assumptions made in connection with the assignment of initial values to the variable parameters of the basic tectonospheric Earth model derived from the dual primeval planet hypothesis.

Since the primordial Earth is fractured into octants by 3 mutually-orthogonal central planes, its surface is encircled by 3 mutually-orthogonal great-circular fracture belts, each having a length of 2 pi times 5400 km = 34,000 km, roughly, or about 3 x 34,000 km = 102,000 km total for the three belts.

The Geometrico-Mechanical Behavior of the Basic 5400-km Primordial Earth.

In order to analyze the geometrico-mechanical behavior of the basic 5400-km primordial Earth, it will be assumed that the basic behavior of the primordial Earth is such that it tends to equilibrate itself to a state of minimum energy. The driving mechanism of the system may be expressed as a function of the disequilibration energy inherent in its initial state. In simplest terms, it consists essentially of the resultant of two factors, each of which has been varying in a predictable manner during the past 4.6 b. y.: (1) the potential energy inherent in the geogenetically disequilibrated shape of the basic 5400-km primordial Earth; and (2) the selectively-channeled energy from the preferential flow of heat, and possibly of volatiles, outward from the three mutually-orthogonal belts which are "active" along the tensile portions of the surficial traces of the 3 planes separating the solid portions of the primordial Earth into 8 octantal subtectonospheric blocks.

When the driving mechanism of the basic model is applied to the primordial Earth, the behavior will depend upon certain geometrico-mechanical constraints inherent in the model. The simpler of these constraints may be summarized in terms of several types of equilibration, or tendency to assume a status of minimum energy:

Rotary Equilibration. Fig. 2-2 shows schematically the expected basic behavior of the original 5400-km primordial Earth in response to the fundamental driving mechanism of the model from the standpoint of rotary equilibration. Basically, there are three degrees of internal rotational freedom, constrained by the configuration of the model to act along the three mutually-orthogonal planes along which the octantal fracturing occurred.

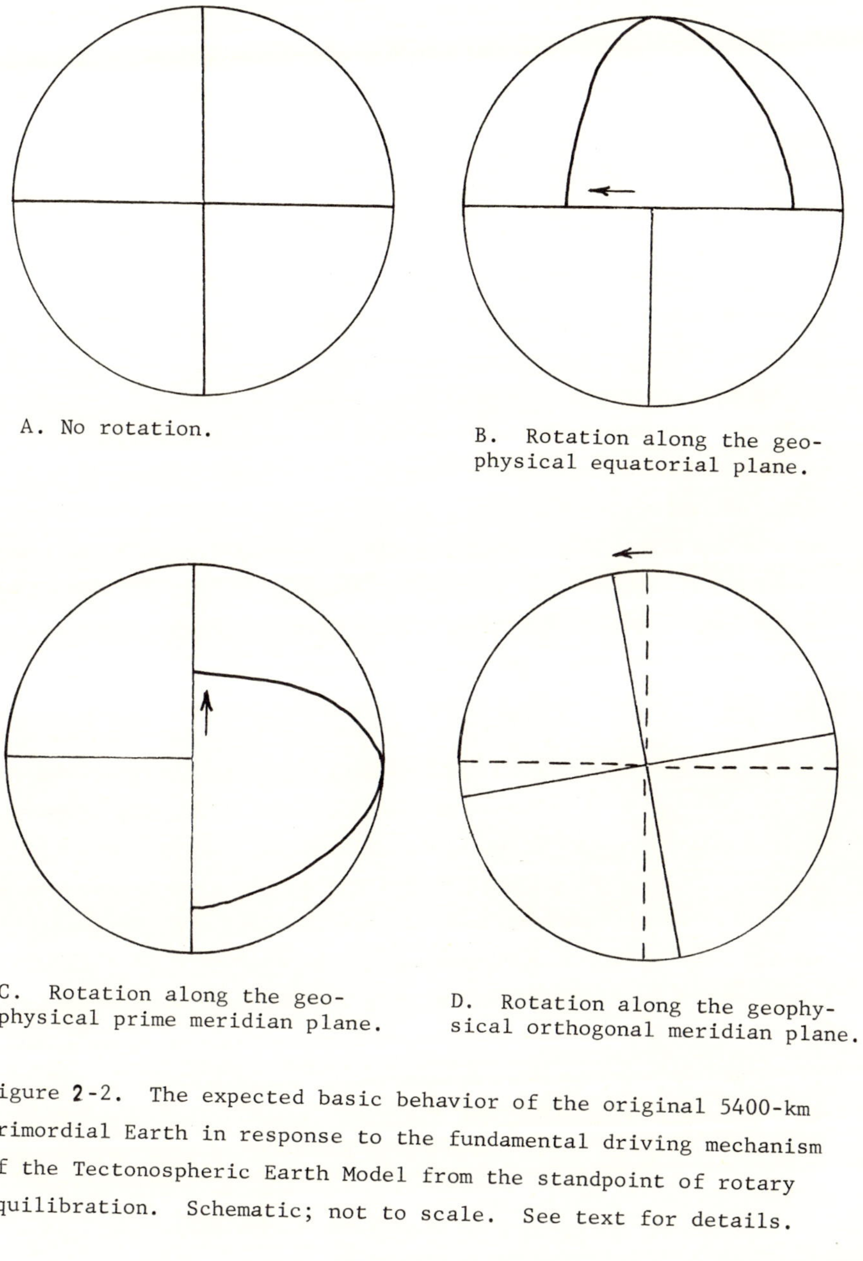

Figure 2-2. The expected basic behavior of the original 5400-km primordial Earth in response to the fundamental driving mechanism of the Tectonospheric Earth Model from the standpoint of rotary equilibration. Schematic; not to scale. See text for details.

It is noticed that, once the rotational motion has begun in any one of the three mutually-orthogonal fracture planes, the gross rotational motion in the other two planes is restricted until a net rotation of n pi/2 shall have transpired in the first plane, where n = 0, 1, 2, 3, - - -.

Radial Translatory Equilibration. Fig. 2-3 shows schematically the expected basic behavior of the original 5400-km primordial Earth in response to the fundamental driving mechanism of the model from the standpoint of radial translatory equilibration. The spherical octants are free to move radially with respect to the center of the model. From the principle of hydrostatic adjustment applied on a global basis, about half of the octants are subsided, at any one time, while the others are elevated with respect to the basic 5400-km radius. The magnitude of the radial motion visualized here is of the order of cm/yr, with maximum amplitudes of the order of ten km.

Transverse Translatory Equilibration. Fig. 2-4 shows schematically the expected basic behavior of the original 5400-km primordial Earth in response to the fundamental driving mechanism of the model from the standpoint of transverse translatory equilibration. Within the constraints of the model, some of the octants (normally two or four) may move transversally along the three mutually-orthogonal fracture-planes. The magnitude of this transverse motion is of the order of cm/yr, with maximum amplitude of the order of km.

Individual Octantal Motion. Fig. 2-5 shows schematically the expected basic independent behavior of the individual octants of the original 5400-km primordial Earth. Within the constraints of the model as a whole, each of the spherical octants

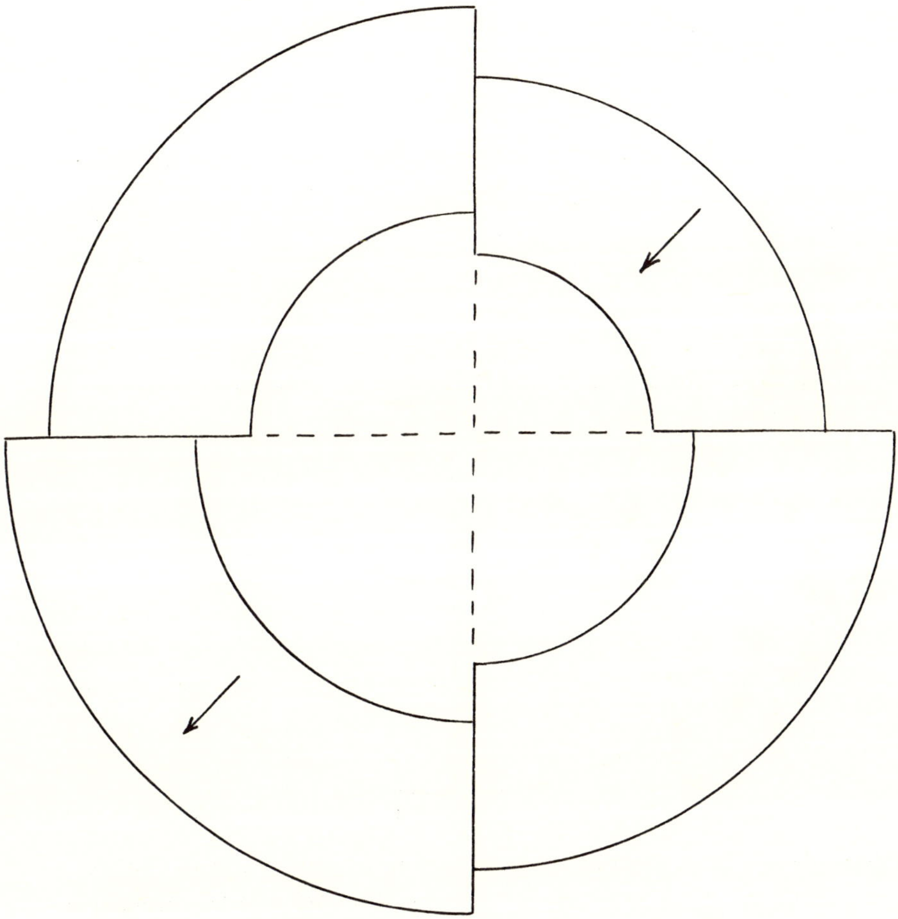

Figure 2-3. The expected basic behavior of the original 5400-km primordial Earth in response to the fundamental driving mechanism of the Tectonospheric Earth Model from the standpoint of radial translatory equilibration. Schematic; not to scale. See text for details.

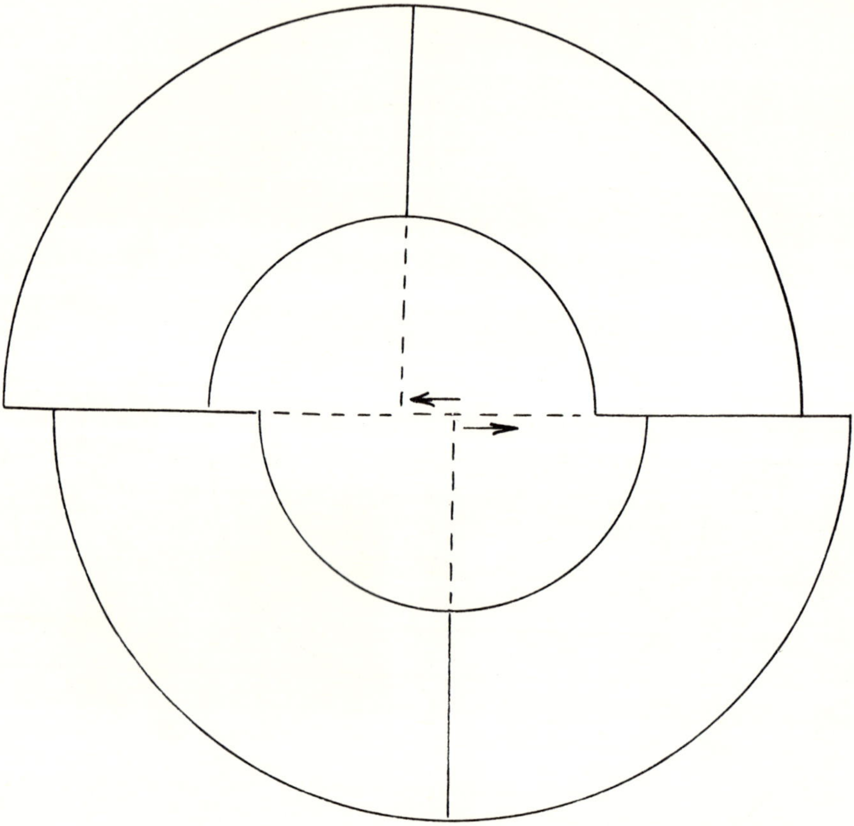

Figure 2-4. The expected basic behavior of the original 5400-km promordial Earth in response to the fundamental driving mechanism of the Tectonospheric Earth Model from the standpoint of transverse translatory equilibration. Schematic; not to scale. See text for details.

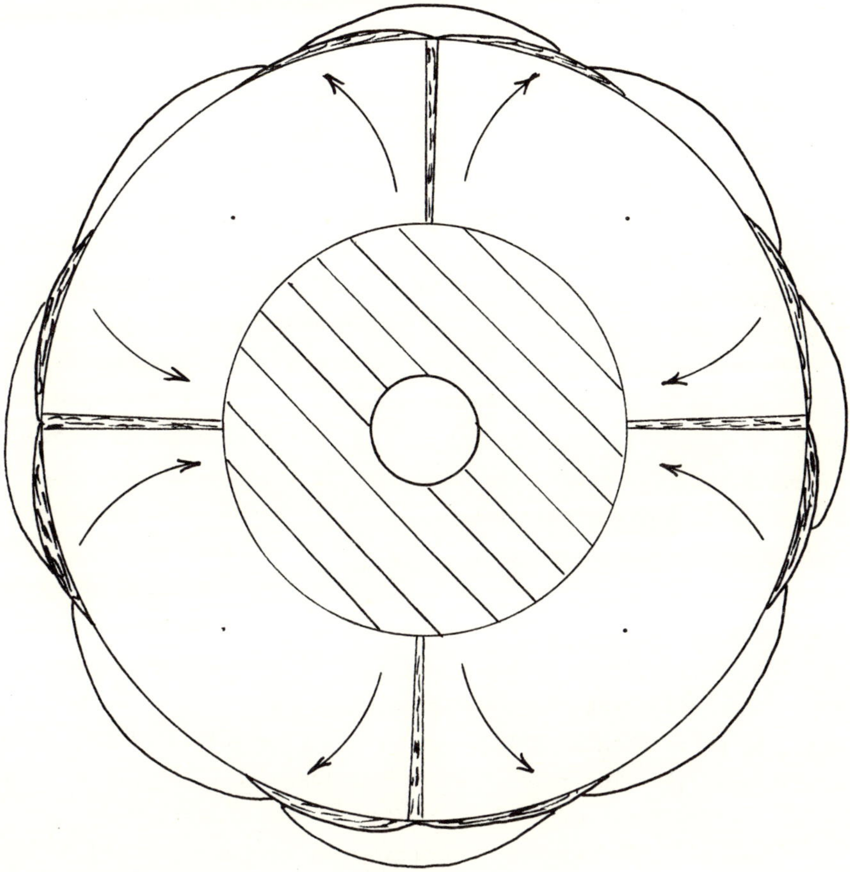

Figure 2-5. The expected basic independent behavior of the individual octants of the original 5400-km primordial Earth in response to the fundamental driving mechanism of the Tectonospheric Earth Model. Schematic; not to scale. See text for details.

has several degrees of freedom with respect to its individual center of gravity. Because of the constraints of the model, this evidences itself mainly as alternate tension and compression along the three mutually-orthogonal fracture planes. This effect is oscillatory, with a long time constant. At any particular time, other things being equal, tension should be found at about half of the octantal contact surfaces with compression at the others. Thus, since each great circle of the basic 5400-km primordial Earth is about 2 pi 5400 = 34000 km long, or 3 x 34,000 = 102,000 km total for all three great circles, then about ½ x 102,000 = 51,000 km of the surficial fracture pattern of the primordial Earth should be in tension while an equal amount is in compression (or neutral). Since the model is continually in the process of equilibrating itself, the locations of the specific segments which are in tension will tend to drift continually along the 102,000-km length of fracture system on the surface of the primordial Earth. Thus, from the standpoint of an external observer, the specific segments in tension would appear to be moving along the 102,000-km linear system in a cyclical but aperiodic manner, since the equilibration of the primordial Earth would, of necessity, be proceeding in a like manner (i. e., cyclically but aperiodically). The rate of such shifting of the tensile segments is of the order of cm/yr.

Segments not in tension at any given time are assumed to be in compression or in a neurtal state. Such segments, whether compressive or neutral, would lie between successive tensile segments and would, therefore, also shift along the 102,000-km subtectonospheric fracture system in a cyclical but aperiodic manner.

Post-Primordial Tectonospheric Accretion.

In evolving the post-primordial Earth from the basic 5400-km primordial Earth, three basic evolutionary modes may be assumed, depending upon the degree to which the 5 terrestrial octants of Earth Prime are fragmentized prior to accretion onto the primordial Earth. These modes are: (1) an octantal-fragment mode; (2) a multiple-fragment mode; and (3) a composite-fragment mode.

The Octantal-Fragment Accretionary Mode.

In the octantal-fragment mode, the 5 octants of Earth Prime are not completely fragmentized prior to accretion onto the basic 5400-km primordial Earth. In this case, parts of each of the 5 octants of Earth Prime might still exist today as unmodified portions of the Earth's tectonosphere, perhaps even as subsurficial portions of presently-existing continental shields. This possibility is considered elsewhere (Tatsch, 1972a: chapter 7).

The post-primordial Earth, in the case of the octantal-fragment mode, evolves essentially from a combination of two processes: (1) the accretion of 5 octants of Earth Prime onto the basic 5400-km primordial Earth; and (2) the differentiation, metamorphism, reworking, igneous activity, and other results of the action of the driving mechanism of the basic model during the past 4.6 billion years (b. y.).

It may be recalled that the driving mechanism of the basic model consists (and has consisted since primordial times) essentially of the resultant of two factors, each of which has been varying in a predictable manner during the past 4.6 b.y.: (1) the potential energy inherent in the geogenetically-disequilibrated shape of the basic 5400-km primordial Earth; and (2) the selectively-channeled energy from the preferential flow of heat, and possibly of volatiles, outward from the three mutually-orthogonal "active" belts formed by the basic octantal fracture pattern of the 5400-km primordial Earth.

Here, as in subsequent discussions and analyses regarding the evolution of the Earth's tectonosphere during the past 4.6 b. y., it is necessary to recall that, in a homogeneous sphere of 6371-km radius, the inertia of the inner 5400-km part normally exeeeds that of the outer 971-km "shell" by at least 60%. Thus, for most purposes, it may be assumed that the disequilibrating inertia of the inner 5400-km part of the model exceeds (and has always exceeded) the equilibrating inertia of the outer 971-km part by a factor of at least 60%. For a non-homogeneous, differentiated body such as the actual Earth, this inertial factor would, of course, normally be even greater than 60%. The significance of this figure is more obvious when some of the mechanics and geometry of the model are discussed in connection with some of the observed behavioral patterns of the actual Earth, such as has been done by the author (Tatsch, 1972a: chapter 8, for example).

The Multiple-Fragment Accretionary Mode.

The multiple-fragment accretionary mode differs from the octantal-fragment mode primarily in the degree to which the 5 "terrestrial" octants of Earth Prime were fragmentized prior to accretion onto the basic 5400-km primordial Earth. For example, if the degree of pre-accretion fragmentation was fairly complete, then it is likely that none of the actual surficial rocks of the present continental shields would now be identifiable with any unmodified portions of the original octants of Earth Prime. This is discussed in greater detail elsewhere (Tatsch, 1972a: chapters 7 and 10).

The Composite-Fragment Accretionary Mode.

The composite-fragment accretionary mode may be considered as an intermediate mode lying between the octantal-fragment mode and the multiple-fragment mode. Depending upon the specific degree of fragmentation assumed for the octants of Earth Prime prior to accretion onto the basic 5400-km primordial Earth, this mode occupies any one of a myriad of intermediate positions within the entire continuum, or envelope, bounded by the octantal-fragment mode and the multiple-fragment mode as extremes.

By extending the limits of the mean, the composite-fragment mode may be made to include one or both of the extremes, thereby collapsing the three modes into two or one, respectively, In this analysis, the three modes will be treated as separate and distinct modes. In the last analysis, any salient differences in the Earth models evolved by the three modes are virtual rather than real; but, by handling the three modes separately, the analysis is more complete and perhaps easier to follow. For this reason, subsequent discussions will assume that there are actually three modes rather than a single mode spread across an entire spectrum, or envelope, including the end points, of the continuum defined by the two extreme modes. The efficacy of this analytical approach is perhaps more obvious when the present continental structures are analyzed in terms of the most likely size of the post-primordial subcratonic blocks that existed within the Earth's tectonopshere, 4.6 b. y. and 3.6 b. y. ago, as well as during the other critical times in the orogenic-cratonic evolution of the Earth's tectonosphere (See, e. g., Tatsch, 1972a: chapters 7 and 15).

<u>The Post-Primordial Earth</u>.

The post-primordial Earth, as derived from the dual primeval planet hypothesis, consists essentially of two parts: (1) an actively-equilibrating, or mobile, 5400-km primordial Earth, differentiated, and fractured basically into octants by three mutually-orthogonal central planes; and (2) an overlying tectonosphere, consisting essentially of the remnants of 5 fragmented octants of Earth Prime accreted onto the 5400-km primordial Earth to a depth of about 1000 km, as modified continually by the driving mechanism of the basic model. Since the primordial Earth loses its identity by becoming incorporated into the post-primordial Earth, it will henceforth be referred to as the Earth's subtectonosphere, or the subtectonospheric portion of the model.

Fig. 2-6 is a schematic sketch of the post-primordial Earth, showing the earliest effects of both igneous and accretionary activity within the tectonosphere. Accretionary activity is indicated schematically at A; igneous activity at I. The octantal blocks of the subtectonosphere are marked S.

Salient Implications of the Basic Tectonospheric Earth Model.

At this point, it is well to list some of the salient implications of the basic Tectonospheric Earth Model in order that the reader may obtain an overall perspective of their nature and scope:

a. Since the 1000-km accretionary material was added to the primordial Earth over a considerable period of time, it follows that various accreted fragments, either individually or in global and regional "shells", would have retained considerable freedom of motion, both with respect to the other accreted fragments and with respect to the basic 5400-km subtectonosphere. Also, it follows that such freedom of motion would most likely be detectable along various planes, surfaces, and shells of preferential rheidity. This is discussed elsewhere (Tatsch, 1972a); there some of these effects are identified as "wave guides", "low-velocity zones", and similar features and phenomena within the tectonosphere. In this connection, it is well to recall that, at a rate of 1 m/yr, a depth of 1000 km is accreted in 1 m. y.; at a rate of 1 cm/yr, it is accreted in 100 million years (m. y.).

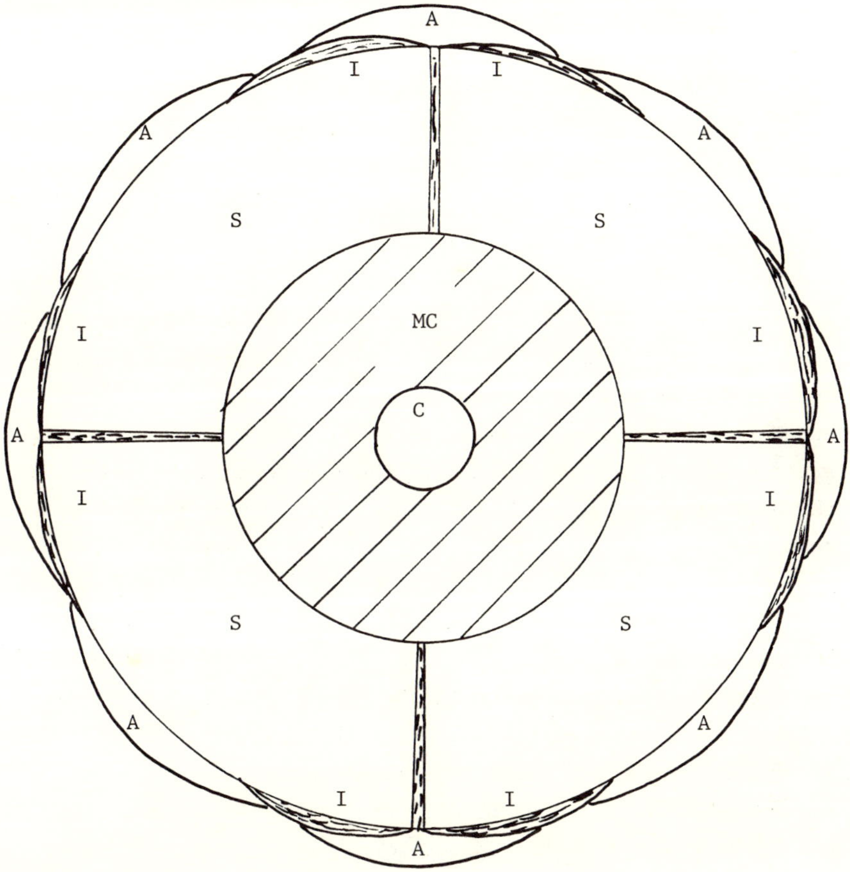

Figure 2-6. The post-primordial Earth, showing the combined effects of primeval igneous and accretionary activities. Accretionary activity is indicated by A; igneous activity by I. The octantal blocks of the subtectonosphere are marked S. MC is the molten core. C is (probably) a small central solid core. Schematic; not to scale. See text for details.

b. The Earth's fracture system consists essentially of the projections, through the tectonosphere, of the basic primordial fracture system of the subtectonosphere. Since the tectonosphere is not homogeneous, such projections are not normally simple, radial projections but somewhat more complex. Other things being equal, the probable error in locating the surficial position of an "energy source" projected from a point on the subtectonosphere through a vertical thickness of 1000 km of average tectonospheric material is roughly 1000 x pi/4, or about 785 km. The significance of this "probable error" figure, as well as the causes for, and magnitudes of, variations in its values are discussed in another work (Tatsch, 1972a; chapter 6), where "cones of activity" are considered in connection with the mechanics and geometry of the evolution of the Earth's tectonosphere during the 4.6 b. y. that the Earth is believed to have been in existence.

c. The Earth's basic deep-fracture system is continuous and worldwide in both the inner (subtectonospheric) and outer (tectonospheric) parts of the model. The latter is basically a consequence of the former, within the constraints of the driving mechanism and the octantally-fractured geometry and mechanics inherent in the model. From this, it appears that the Earth's deep trenches should form a global pattern identifiable with the three mutually-orthogonal great-circular belts of the subtectonosphere. This is discussed elsewhere (Tatsch, 1972a: chapters 6, 7, 8, and 18).

d. "Active" regions (new islands, new mountains, seismic ridges, new volcanoes, etc.) lie roughly above the basic fracture planes of the inner, subtectonospheric part of the model. Due to the effects of planes, surfaces, zones, and shells of preferential rheidity and heat flow, these "active" regions would not, of course, be expected to lie directly above the subtectonospheric fracture system (Tatsch, 1972a: chapter 9).

e. "Inactive" or "fossil" regions (older mountains, older islands, older volcanic belts, seamounts, aseismic ridges, older mineralization belts, etc.) are identifiable with former positions of projections from the subtectonospheric fracture system. To an external observer, these fossil, or inactive, regions appear to have "drifted", in time, with respect to the presently active regions. Normally, such "drifting" involves both translation and rotation, where neither is necessarily linear with respect to time. Also, it might be more accurate to say that the subtectonospheric fracture system, rather than the surficial manifestations thereof, have "drifted", or perhaps even more accurately that both the surficial manifestations and the basic subtectonospheric fracture system have drifted, depending upon whether one wishes to assume a fixed or mobile datum during the analysis. In either case, of course, the results are the same. Also, according to the basic model, the "fossil" features are spatially subparallel to, and temporally derivable from, the "active", or present, features by backward extrapolation procedures. Needless to say, such spatial and temporal relationships are not necessarily linear with respect to time. This is discussed in greater detail elsewhere (Tatsch, 1972a: chapters 7, 8, and 9).

f. Other things being equal, the age of any particular surficial manifestation of the Earth's internal behavior (such as a mountain range, a crustal rift-ridge system, a volcanic belt, etc.) may be approximated roughly by measuring the <u>gross</u> distance by which the particular surficial manifestation has "drifted" from the present position of its genetic subtectonospheric fracture plane. Thus, it is not too surprising that the Appalachian-Caledonian (Upper Paleozoic) orogeny is geographically removed about 30° from the surficial trace of the nearest sub-tectonospheric fracture plane (i. e., Galapagos-Gibraltar), because a drift of 1 cm/yr is roughly equivalent to about 30° per 300 m. y. This is discussed in greater detail elsewhere (Tatsch, 1972a: chapters 7, 8, 9, 10, 12, 13, 16, 17, 18, and 20).

g. The oldest geological features (or surficial manifestations of the Earth's internal behavior), such as Precambrian shields, presently occupy positions identifiable with the basic fracture system of the subtectonosphere. Specifically, they are either: (1) identifiable with the centers of the surficial octants of the subtectonosphere, if stable; or (2) identifiable with the edges of the surficial octants of the subtectonosphere, if they are unstable, active, or in the process of rifting. An example of the latter might be the South American-African "shield" 200 m. y. ago, which has since been rifted and separated as shown schematically in Fig. 2 -7. A more-detailed discussion of this is contained elsewhere (Tatsch, 1972a: chapter 17).

h. Likewise, the newest surficial features of the Earth's internal behavior would be expected to be located near the surficial traces of the three mutually-orthogonal subtectonospheric fracture planes projected to the surface of the Earth.

i. One of the several "unbalanced" conditions existing in the post-primordial Earth resulted from the fact that portions of the Earth's surface must, at various times during the history of the post-primordial Earth, have become squarely juxtaposed above the surficial manifestations of the tectonospheric fracture system. Among the consequences of such a situation would be

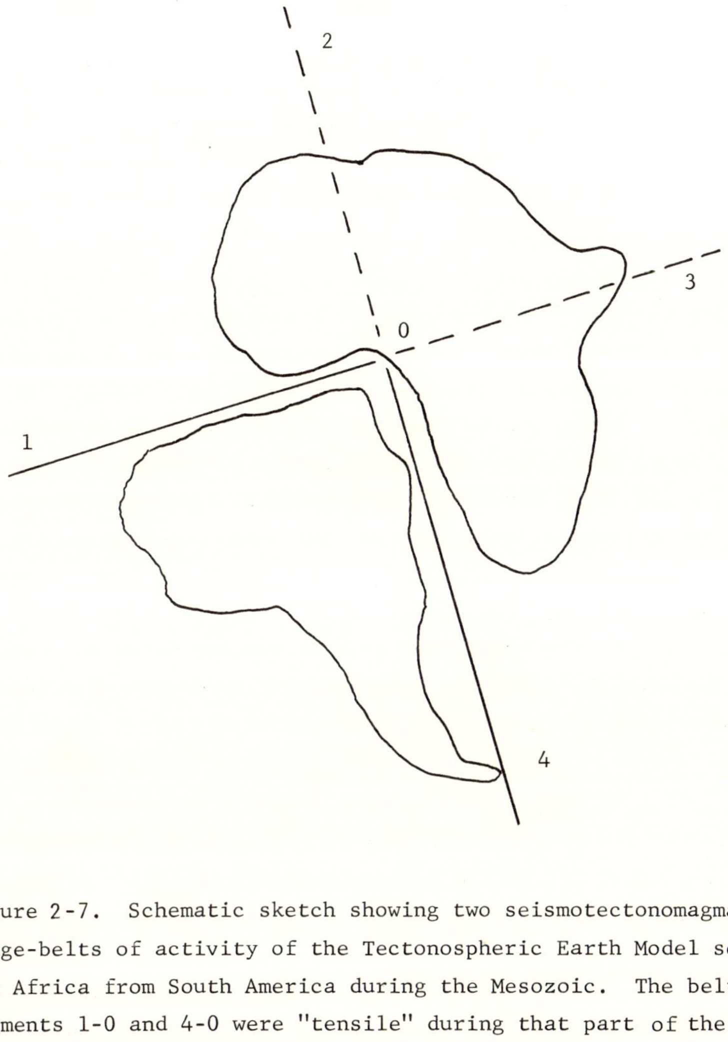

Figure 2-7. Schematic sketch showing two seismotectonomagmatic wedge-belts of activity of the Tectonospheric Earth Model severing Africa from South America during the Mesozoic. The belt segments 1-0 and 4-0 were "tensile" during that part of the Mesozoic; the segments 2-0 and 3-0 were compressive or neutral. See text for details. (Adapted from Tatsch, 1972a).

sea-floor spreading and continental drift (Tatsch, 1972a: chapters 16 and 17).

j. The stress release and other equilibrating effects from the potential energy inherent in the geogenetically-disequilibrated shape of the subtectonosphere are the basic cause of the Earth's deep seismicity (Tatsch, 1972a: chapter 8).

k. Mantle convection may be motivated by the preferential flow of heat along the patterns provided by the subtectonospheric fracture system. Under this concept of the model, mantle convection is a manifestation of the driving mechanism of the basic model. Thus, unlike most other hypotheses, the Tectonospheric Earth Model does not invoke mantle convection as a motive force for sea-floor spreading and continental drift. Rather, mantle convection is derivable from the basic model as an inherent manifestation of the same driving mechanism that drives sea-floor spreading, continental drift, and other omniductive phenomena on the surface of the Earth. The mechanics and geometry of these phenomena, as viewed by the model, are discussed in another work (Tatsch, 1972a: chapters 16 and 18).

l. Igneous activity, both intrusive and extrusive, is motivated by the preferential flow of heat, and possibly of volatiles, outward from the subtectonospheric fracture system (Tatsch, 1972a: chapter 10; 1973a: chapter 4).

m. The preferential flow of heat outward from the subtectonospheric fracture system may be employed to furnish the basic motive power for, and a stable framework for maintaining, a geomagnetic dynamo with a long time constant (Tatsch, 1972a: chapter 14).

n. The roughly antipodal locations of the oceans with respect to the continents is basically a consequence of two factors: (1) the radial translatory motion of the subtectonospheric octantal blocks with respect to the center of the model; and (2) hydrostatic adjustment, or isostasy, on a global scale over a long period of time (Tatsch, 1972a: chapter 15).

o. If viewed on a global scale, certain features such as fault lines, rift-ridges, mountain belts, seamount chains, etc., should normally tend to display "parallel" surficial patterns, intersected by transverse features. As a result, other things being equal, they should tend to form surficial manifestations suggestive of pinnate, grid, and arcuate patterns. These reflect both translation and rotation between the subtectonospheric blocks and the surface of the Earth, most probably along planes, shells, surfaces, and zones of preferential rheidity within the tectonosphere (Tatsch, 1972a: chapters 6 and 18).

The above are a few of the "first order" implications of the basic Tectonospheric Earth Model. The details of these and other implications of the model are discussed elsewhere (Tatsch, 1972a). Higher order implications of the model are the subject of other publications (See, e. g., Tatsch, 1973a through 1973j; 1974b; 1975a; 1975b; 1975c).

<u>Global Surficial Manifestations of the Earth's Internal Behavior Predictable from the Basic Tectonospheric Earth Model</u>.

One of the postulates of the basic Tectonospheric Earth Model is that almost all surficial features and phenomena of the Earth, both past and present, are fundamental consequences of a single, deep-seated driving, or causal, mechanism that has resided within the Earth since the formation of the primordial

Earth, about 4.6 b. y. ago. The details of exactly how such surficial manifestations evolved as a consequence of the behavior of the Earth's tectonosphere during the past 4.6 b. y. is beyond the scope of this book, but many of these details have been discussed elsewhere (See, e. g., Tatsch, 1972a). It may be well to list some of these surficial manifestations of the Earth's internal behavior, in order that the reader may have an appreciation for the general nature and scope of those surficial manifestations that, according to the model, are internally-induced in a predictable manner. These include: global and regional tectonism and orogenesis; sea-floor spreading and continental rifting; global and regional heat-flow patterns; the Earth's magnetic field and its variations, including cyclical but aperiodic reversals in polarity; the Earth's gravity field; volcanic activity; mountain building and related geosynclinal activity; the development of continental shields, mobile belts, fracture systems, rifts, and ridges; continental splitting, separation, and drifting; earthquakes and global and regional seismicity patterns; gravity anomalies and global and regional geoidal patterns; heat-flow anomalies and global and regional thermal patterns; electrical conductivity anomalies; global seismotectonomagmatic and mineralization belts; the Earth's plate-tectonic and omniductive behavior; etc.

At this point it is well to examine one of the surficial manifestations of the Earth's internal behavior, in an attempt to provide a tentative answer to the compound question: Is there any observational evidence that the Earth might, in fact, be representable by the Tectonospheric Earth Model; and, if so, how might the present positions of the eight hypothetical subtectonospheric blocks be identified or located?

To answer this compound question, consideration may be given to a preliminary analysis of any one of several sets of applicable observational data regarding the gross surficial characteristics of the Earth. For example, one such set of data is that showing the locations of the active volcanoes. Since volcanoes appear to be thermally induced and since they are surficial manifestations of tectonospheric activity of some type, they might be expected, according to the model, to lie along segments of global belts defined roughly by the surficial traces of the projections from the three mutually-orthogonal great-circular belts forming the subtectonospheric fracture system.

Therefore, as a first approximation to the location of the surficial manifestations of the subtectonospheric fracture system, we may perform a three-dimensional least-squares analysis of the distribution of the Earth's active volcanoes. Allowing for the probable error expected from the 1000-km projection through the heterogeneous tectonospheric material, we should expect almost all of the Earth's active volcanoes to lie within a surficial distance of about 1000 x pi/4 $\doteq$ 785 km from segments of the radial projections of the subtectonospheric fracture pattern.

When the suggested theee-dimensional least-squares analysis is done, it is found (Tatsch, 1964a) that approximately 93% of the active volcanoes fall within the belts predicted by the model. Specifically, the active volcanoes appear to form segments of three great-circular belts intersecting at the following 6 points, which may be designated tentatively as the "basic tectonospheric points" of the present surficial manifestitions of the basic Tectonospheric Earth Model: 55°N, 165°W (Aleutians); 55°S, 15°E (Bouvet); 5°S, 85°W (Galapagos); 30°N, 5°E (Gibraltar); 5°N, 95°E (Bengal); and 30°S, 175°W (Kermadecs).

Because the probable error of the projection from the subtectonosphere to the surface of the Earth is about 785/111 $\doteq 7^{o}$, no purpose is served in attempting to locate the coordinates of the tectonospheric points more "accurately" than about 5° in either latitude or longitude. The geographic names (Aleutians, Bouvet, etc.) that are shown above in parentheses are intended for associative identification of the general areas of the points, rather than as exact geographical descriptions of the points. The associative names will normally be used in subsequent discussions as a matter of convenience; but it should be understood that these points are projections of 6 other points that form a pattern on the surface of the subtectonosphere, 1000 km beneath the surface of the Earth.

For the sake of simplicity, the first two points (Aleutians and Bouvet) may be referred to as the "tectonospheric poles" of the present surficial manifestations of the basic Tectonospheric Earth Model; the other four points (Galapagos, Gibraltar, Bengal, and Kermadecs) may be considered as defining roughly the "tectonospheric equator" of the present surficial manifestations of the basic Tectonospheric Earth Model. Fig. 2-8 shows the above tentative "tectonospheric coordinate system" of the present surficial manifestations of the basic Tectonospheric Earth Model superimposed upon a Mollweide projection of the salient geographical features of the actual Earth. In the figure, the three mutually-orthogonal great circles (ACDFA, ABDEA, and CBFEC) are the radial projections of three similar great circles defining the subtectonospheric fracture system. Each of the intersections at A, B, C, D, E, and F is an orthogonal intersection, and each of the 8 triangles (CBA, CBD, CED, CAE, FEA, FDE, FBD, and FBA) is an equilateral spherical triangle (i. e., with all sides and angles equal to 90°). On the surface of the tectonosphere, 90° equals to about 10,000 km. On the surface of the subtectonosphere, 90° equals about 8500 km.

The great circle, ABDEA, also marked QQQQ in Fig. 2-8, is a projection of the subtectonospheric equator and it defines roughly the tectonospheric equator of the present surficial manifestations

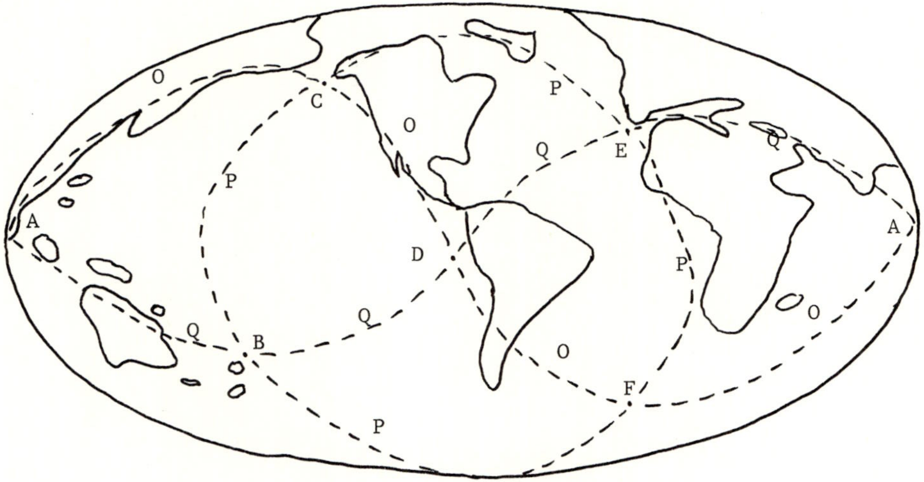

Figure 2-8. Tentative "tectonospheric coordinate system" of the present surficial manifestations of the basic Tectonospheric Earth Model superimposed upon a Mollweide projection of the salient geographical features of the actual Earth. See text for explanation of symbols.

of the basic Tectonospheric Earth Model. The great circle, CBFEC, also marked PPPP, is a projection of the subtectonospheric prime meridian and it defines roughly the tectonospheric prime meridian of the present surficial manifestations of the basic Tectonospheric Earth Model. Similarly, the great circle, ACDFA, also marked OOOO, is a projection of the subtectonospheric orthogonal meridian.

A tectonospheric coordinate system based on the above is particularly useful as a base for performing harmonic analyses of the Earth's surficial manifestations. According to the dual primeval planet hypothesis, the Earth's surficial manifestations are tectonospherically induced and, therefore, "harmonic" with respect to the tectonospheric coordinate system, rather than with respect to the geographic coordinate system normally used for harmonic analyses when using other models.

The Tectonospheric Earth Model as a Supplementary Tool in the Exploration for Minerals.

Briefly stated, the Tectonospheric-Earth-Model approach to mineral exploration is based upon a straightforward premise: if one knew today where all the Earth's minerals were 4.6 b. y. ago, and if he could define the geometrical, mechanical, thermal, and chemical behavior of the Earth during the past 4.6 b. y., then he should be able to predict the present locations of the Earth's mineral deposits (Tatsch, 1973a).

The following chapters discuss how the Tectonospheric Earth Model may be used to define the complex geometrical, mechanical, thermal, and chemical interrelations that exist among igneous activity, folding, faulting, uplifting, subsidence, and other omniductive geophenomena that have contributed to the origin, evolution, and present characteristics of copper deposits.

Chapter 3

THE GEOMETRICAL, MECHANICAL, THERMAL, AND CHEMICAL BEHAVIOR OF THE EARTH DURING THE PAST 4.6 BILLION YEARS

Geophysical exploration and deep drilling are the clue to hidden copper deposits and to covered structures. But geophysical data and deep-drilling results can be properly interpreted only if we have some idea of how the Earth's substructure has originated, how it has changed during the past 4.6 b. y., and how it is behaving now.

Global-tectonic and related concepts have provided many answers regarding the geometrical, mechanical, thermal, and chemical aspects of the Earth's internal behavior that have contributed to the origin, evolution, and present characteristics of copper deposits (See, e. g., Coleman, 1975; Nikolayev and Tregub, 1970; Ahrens, 1973; Tarling and Runcorn, 1973; Turcotte and Oxburgh, 1972; Watson, 1973; Baker et al., 1972; Boucot, 1972; Zen, 1972; Kurita, 1973; Bowen, 1973; Pettijohn, 1970; Jackson, 1973; Kennedy, 1973; Steinhorn, 1973, York and Helmberger, 1973; and earlier references cited in Tatsch, 1972a: chapters 16, 17, and 18).

The Tectonospheric Earth Model supplements the basic global-tectonic and related concepts by (1) extending the applicable spatio-temporal domain to a depth of at least 1000 km and to a span of about 4.6 b. y.; and (2) providing a single, long-lived, deep-seated, global driving mechanism that has been motivating the Earth's internal behavior during the 4.6 b. y. that the Earth is believed to have been in existence. It is well to consider, at this point, some of the aspects of these two supplementary extensions that have been provided by this model. This may be done by summarizing some of the results of using the Tectonospheric Earth Model to explain the geometrical, mechanical, thermal, and chemical behavior of the Earth during the 4.6 b. y. that the Earth is believed to have been in existence.

Armed with sophisticated sensors, instruments, and computers, a modern scientist can measure and otherwise observe many surficial features, phenomena, and manifestations of the Earth's internal behavior. He is not always sure just what causal relationships exist between what he observes and what is happening inside the Earth. He can, for example, see the crust rifting at certain places; and, when this occurs, he calls it "sea-floor spreading"; but he has not yet been able to completely understand just what sea-floor spreading is or what causes it. Nor does he yet understand precisely why sea-floor spreading occurs at certain places but not at others (Gilluly, 1971; Dickinson, 1973a; Coney, 1973a, 1973b; Dalrymple et al., 1973; Cloud, 1974; White, 1972; Thompson and Melson, 1972; Milanovsky, 1972; Heirtzler, 1972; Meyerhoff and Meyerhoff, 1970; Meyerhoff and Teichert, 1971; Oxburgh, 1972; Lowry, 1974; and Watts and Talwani, 1975).

In a similar manner, when he sees lithospheric plates moving about the surface of the Earth, he computes their instantaneous centers of rotation and calls this phenomenon "plate tectonics", but he has not been able to completely comprehand precisely what plate-tectonics is or what causes it. Nor does he yet understand exactly why the lithospheric plates form into pieces of certain sizes and shapes rather than others. He suspects that the cause lies deep below, but he is not sure just where, nor what the causal mechanism is or how it operates to break the lithosphere into a few large plates and many small subsidiary and peripheral plates (See, e. g., Lachenburch and Thompson, 1972; Osmaston, 1973; Smith, 1972; Morgan, 1972; Termier and Termier, 1973; Lowry, 1974;

Radulescu and Sandulescu, 1973; Nelson and Temple, 1972; McKenzie, 1972; McElhinny, 1972; Engels and Kelm, 1972; and Burk, 1972).

In order to see how the present plates might have looked in "pre-drift" reconstructions, various attempts have been made to reassemble these into supercontinents, such as Pangea, Gondwana, and Laurasia (See, e. g., Bullard et al., 1965; Dietz et al., 1970; Drake et al., 1968; Hallam, 1971; and Schenk, 1971). More recently, Dalziel and Elliot (1971), Roper (1974), and Walper and Rowett (1972), among others, have attempted to eliminate some of the "fitting" problems that were encountered in attempting to fit all the present continents into a unified Paleozoic configuration. The Dalziel-Elliot (1971) model addresses the problem encountered by the earlier models in fitting South America, the Scotia Arc, and west Antarctica. The Roper (1974) and the Walper-Rowett (1972) models considered the bothersome area in previous fits. An analysis by the author and his associates (See, e. g., Tatsch, 1972a; Tatsch and Talbot, 1971) suggested that there is no such thing as a single predrift reconstruction, because the continents have been drifting ever since they first formed. This led to the question of what caused the lithosphere to break up at various times. Closer scrutiny shows that the break-up of the lithosphere is a surficial manifestation of the Earth's internal behavior. But what, precisely, is the nature of the Earth's internal behavior that causes this and other surficial manifestations?

For ease of presentation, three aspects of the global surficial features and phenomena may be considered: (1) a global view of the Earth's present surficial features; (2) the most probable surficial features of the proto-Earth 4.6 b. y. ago; and (3) the evolution of the Earth's present surficial features from those of the proto-Earth during the past 4.6 b. y. Since the evolutionary period appears to have been interrupted about 3.6 b. y. ago, the period will be considered in two parts: 4.6 b.y. to 3.6 b. y. ago, and 3.6 b. y. ago to the present.

Orogenic-Cratonic Structure of the Continents.

Scientists have long suspected that the orogenic-cratonic structure of all continents is more-or-less the same and that, in the simplest terms, continents have grown by "accretion" onto a central shield or nucleus. Exactly what caused them to follow such a pattern of quasi-organized accretionary growth is not known. Besides, most modern hypotheses do not extend back as far as the 3.6 b. y. during which observational evidence supports such growth-from-the-center evolution for the continents.

In viewing the orogenic-cratonic structure of the continents on a global scale, it appears: (a) that orogenic belts are long and rectilinear, i. e., that the accretionary growth may occur along segments of great circles, which unlike small circles plot on the surface of the globe as rectilinear features; (b) that the continents are old and appear to have formed by accretion from the center outward, beginning at least 3.6 b. y. ago; (c) that the oldest orogenic belts are nearest the centers of the continental nuclei, but that farther from the centers there may be a certain amount of overprinting by orogenic belts of different ages; (d) that generally speaking orogenic belts appear to have been active on a worldwide basis in a cyclical but aperiodic manner during at least the past 3.6 b. y.; (e) that the concentration of radioactive elements in continental rocks suggests chemical differentiation of the mantle, or an equivalent process; and (f) that the possibility of such differentiation suggests: (1) that the entire continental areas may be of ultimate mantle origin, and (2) that the primordial Earth might have been completely devoid of crust or continents as we know them today.

This orogenic-cratonic structure is fairly well preserved in certain continents (See, e. g., Bowen, 1974; Burchfiel and Davis,

1972; Churkin, 1972, 1973; Coney, 1972; Davis, 1973; Dawes, 1973; Deutsch and Rao, 1973; Douglas, 1969; Drake, 1970; Gabrielse, 1972; Goodwin, 1973; Grow, 1973; Heezen et al., 1973; Herrin, 1972; Hurley, 1969, 1973; Hurley and Rand, 1969; Kamen-Kaye, 1970; Karig, 1972; Kent and Warman, 1972; Khain, 1972; Krogh and Davis, 1972; Martin, 1969; Menard, 1973; Monger et al., 1972; Moore, 1973; Nairn, 1973; Natland, 1974; Piper, 1973a, 1973b; Roberts, 1972; Robertson, 1972; Roeder, 1973; Scheidegger, 1972; Shackleton, 1973; Silbering, 1973; Smith, 1973; Sugimura and Uyeda, 1973; Velentine and Moore, 1972; Wiggins and Helmsberger, 1973; Wilson, 1972; Wynne-Edwards, 1969).

What is seen, in general, is an old Precambrian shield successively surrounded by long rectilinear orogenic, geosynclinal, or seismotectonomagmatic belts, often broken or otherwise modified by overprinting by subsequent seismotectonomagmatic activity, with the last three orogenies fairly well preserved in at least some continents; e. g., the Appalachian, Innuitian, and Cordilleran orogens of the Canadian shield. Unanswered are questions as to why the orogenic belts are rectilinear and of different ages within any one continent but appear to extend worldwide, with their activity having occurred in cyclical but aperiodic episodes for at least the past 3.6 b. y., and with concentrations of this activity having occurred at about 0.9-b. y. intervals. Also not understood is why the spatio-temporal relationships within all orogenic belts are remarkably similar, regardless of where found or to which period they belong.

When they are considered to be surficial manifestations of the Earth's internal behavior, the orogenic-cratonic structures of the continents suggest a remarkably orderly behavior over an extremely long period of time, perhaps as long as 3.6 b. y. or even 4. 6. b. y.

The Earth's Seismicity and Earthquakes.

It is universally recognized that earthquakes are the result of fracturing occurring at various depths of the crust and upper mantle to depths of at least 800 km. Many of the faults associated with earthquakes do not extend to the Earth's surface, and the epicentral areas of even severe earthquakes provide no evidence of the substructural faults and fractures. In such cases, the only surficial manifestations of the Earth's internal behavior is the seismic shock itself.

Other studies about earthquakes indicate that the causal relationships are not the same in regions of different geological structure and history, such as young folded regions, young platforms, platforms of any age involved in intensive tectonic movements during recent times, and rift zones within continents (See, e. g., Drake, 1970; McBirney, 1969; Barazangi and Dorman, 1969; Petrushevsky, 1969; McGarr, 1969; Molnar and Oliver, 1969; Oliver et al., 1969; Raitt et al., 1969; Balakina et al., 1969; York and Helmberger, 1973; Subbotin et al., 1972; Hales and Herrin, 1972; Griggs, 1972; Kelleher et al., 1973; Huestis et al., 1973).

When viewed on a global scale, in terms of ordinary analyses, neither the Earth's seismicity nor the distribution of earthquakes appears to form an orderly pattern. However, by using special techniques, much progress has been made in recent years to develop the specifications of an unambiguous worldwide earthquake "mechanism". Many observers report evidence indicating that the pressure axis is perpendicular to the trend of the continental seismic belts and that the tension axis is perpendicular to mid-oceanic belts, when analyzed in terms of the earthquake energy-release mechanism (See, e. g., Balakina et al., 1969). However, such evidence tends to be somewhat misleading because some of the so-called "mid-ocean" belts lie within continental areas (i. e., in North America, Central America, Africa, Iceland, etc.). It does suggest, however, that the earthquake energy-release mechanism is truly global and independent of the present positions of the continents and oceans. Thus, considered as surficial manifestations of the Earth's internal behavior, earthquakes suggest that the Earth's internal behavior is controlled by a global mechanism completely independent of continental margins, island arcs, and similar features, except insofar as they may also be surficial features of the same mechanism.

Heat Flow and Other Geothermal Activity.

A detailed description of the Earth's thermal history is, of necessity, a rough guess, because the distribution of heat sources within the Earth is not known with any certainty. However, the following comments may be made in connection with surficial manifestations of the Earth's internal thermal hehavior on a global scale: (a) thermal Earth models based on spherical symmetry are not consistent with the assumption that radioactive elements are more highly concentrated in continental than in oceanic crust; (b) the distribution of radioactive material below the M discontinuity is usually assumed to be somewhat different under oceans than under continents; (c) in a body the size and composition of the Earth, thermal diffusion is so slow that little heat could have escaped from depths greater than about 750 km during the past 4.6 b. y. (Tikhonev et al., 1969; Stacey, 1969), unless it has escaped along paths of preferential heat flow (Tatsch, 1972a: chapter 9); (d) it is often assumed that "for any reasonable distribution" of radioactivity in an oceanic Earth model, the actual effective thermal conductivity at depth must be substantially greater than the average value of thermal conductivity of igneous rocks; otherwise the entire mantle would be molten; (e) thermal conduction is usually presumed to be confined to the upper 700 km of the Earth; and (f) there is evidence that the thermal conductivity of the mantle maintains a temperature profile in equilibrium with internal heat generation (Stacey, 1969).

Although the outward flow of heat through the crust is the only directly measureable surficial manifestation of the Earth's internal thermal behavior, certain tentative conclusions may be drawn from the present-day analyses made in connection with heat-flow data: (a) there is a general correlation of heat flow with geology, the higher heat flows being associated with the more recent orogeny; (b) in specific regions, such as the mid-Atlantic ridge, for example, heat flow values vary considerably, such as is also true for volcanic areas, of course; (c) high heat flow appears to correlate with low gravity (Lee and MacDonald, 1963); (d) continental and oceanic heat flow averages are roughly the same; (e) the features of the geoid do not appear to correlate with the present locations of the continents and oceans; (f) the oceanic mantle is neither hotter nor cooler, on average, than is the continental mantle; (g) the equality of average outward flow of heat from continents and oceans may be the result of a dynamic balance in a convecting (or otherwise well-behaved) upper mantle (Elsasser, 1967); (h) although radioactivity is approximately the same beneath the oceans and continents, there appear to be some differences within the upper mantle (Polyak and Smirnov, 1968); (i) although the global correlation between heat and gravity is not understood, the lack of a global spatial correlation may be the result of a temporal lag between the Earth's internal thermal behavior and the internal behavior producing changes in gravity, perhaps amounting to a temporal lag of millions of years; (j) a more definite correlation at the present time may exist between heat flow and age of basement rock (See, e. g., Lubimova, 1969), because heat flow values appear very uniform in all shield areas (i. e., 0.92 ± 0.38 hfu) and in all Meso-Cenozoic orogenic areas (i. e., 1.92 ± 0.49 hfu), the general correlation being that heat flow decreases with increase in age of basement; (k) preliminary analyses show that heat-flow data correlate with other geophenomena and features, provided that the low "signal" of the system (i. e., the paucity of heat-flow data) can be offset by decreasing the "noise" of the system (such as by the transformation from geographical coordinates to geotectonospheric coordinates, as suggested by Tatsch, 1972a: chapter 3); and (l) temporal variations in heat-flow rates are significantly different for regions of tectonogenesis of dissimilar ages if the deepest levels of the upper mantle are assumed to be responsible for the formation of the tectonic zones and if the tectonogenesis of dissimilar age is characterized by various intensities in the inflow of sialic material from the mantle (Lubimova, 1969).

Thus, when viewed on a global scale, it appears that the Earth's internal behavior reveals itself through thermal manifestations on the surface; but the exact correlation between internal behavior and surficial manifestations of the thermal Earth is not clearly understood, perhaps because of the paucity of global heat-flow data (Roy et al., 1972; Nitsan, 1973). If a generalization may be made, it is that regions of modern orogenesis correlate with high heat-flow rates (See, e. g., Dewey and Burke, 1973; Molnar and Atwater, 1973; Morgan, 1971, 1972; Scrutton, 1973; Shaw, 1973; Shaw and Jackson, 1973; Wilson, 1963).

Fixed versus Unfixed Causal Mechanisms for the Surficial Manifestations of the Earth's Internal Thermal Behavior.

It has been known for some time that many of the Earth's surficial features and phenomena appear to be related to certain aspects of the Earth's internal behavior. One of these relates to the surficial manifestations of the Earth's internal thermal behavior. These surficial manifestations have been analyzed, in the literature and at scientific meetings, in terms of 2 basic concepts:

Concept A. "Paths of preferential heat flow" have been discussed in terms of "wedge-belts of activity" by Tatsch (1963 et seq.).

Concept B. "Hot spots" (and their variants, such as "mantle plumes" and "melting spots") have been discussed by Wilson (1963), Morgan (1971), Jackson et al. (1972), Clague and Jarrard (1973), Dalrymple et al. (1973), Larson and Chase (1970), Molnar and Atwater (1973), Schalanger et al. (1974), and many others.

The primary differences and similarities in these two basic concepts may be summarized:

a. Both concepts are based on the same observational evidence, i. e., the surficial manifestations of the Earth's internal thermal behavior.

b. The causal mechanism is more clearly (if somewhat more speculatively) defined in Concept A (paths of preferential heat flow) than it is in Concept B (hot spots, etc.).

c. The causal mechanism for Concept B has been analyzed in terms of "mantle convection".

d. The causal mechanism for Concept A has been analyzed in terms of the most probable geometrical, mechanical, thermal, and chemical aspects of the Earth's behavior, to a depth of 1000 km, during the 4.6 b. y. that the Earth is believed to have been in existence (Tatsch, 1963 et seq., particularly 1972a).

e. Concept B does not define the spatio-temporal relationships among the various "hot spots", "mantle plumes", "melting spots", etc. That is, there is no agreement within the literature on the extent to which these phenomena are fixed (or moving) with respect to each other or to a "fixed" reference within the Earth's substructure (See, e. g., Silver, 1974; Molnar and Atwater, 1973).

f. Concept A defines the spatio-temporal relationships among the various surficial manifestations of the Earth's internal thermal behavior in terms of the 6 "basic points" of the subtectonospheric fracture system (Tatsch, 1963 et seq., particularly 1972a).

g. Concept B does not clearly identify the source of the magma that motivates that concept. That is, it leaves unspecified whether the source of the magma is (1) a rising plume caused by processes acting deep within the mantle (Morgan, 1972); (2) the local velocity gradients in the asthenosphere caused by the movements of epi-asthenospheric lithospheric plates themselves (Shaw, 1973); (3) the downwelling of "gravitational anchors" (Shaw and Jackson, 1973); or (4) a combination of these and other (undefined) causes.

h. It is difficult to visualize, under Concept B, how a simple mantle plume could account for the near-simultaneous, large-volume volcanism and basalt flows that have been observed along lengthy, narrow rifts in many parts of the Earth's surface during much of the Earth's existence (See, e. g., Natland, 1974). Under Concept A, this difficulty is not encountered (See, e. g., Tatsch, 1972a: chapter 10).

i. Concept A defines the source of the magma in terms of the predictions of the basic Tectonospheric Earth Model (Tatsch, 1963a, et seq., particularly 1972a) and as evidenced in the origin, evolution, and present characteristics of the Earth's mineral deposits (Tatsch, 1973a).

j. The Easter Island seamount chain was once cited as having resulted from a "hot spot" at the East Pacifc Rise. Closer scrutiny of this interpretation, under Concept A, reveals the following:

(1). This interpretation is not consistent with either of the two "spreading" models that have been proposed for that region (Harrison et al., 1974): (a) a "jumping" ridge crest, or (b) the presence of two spreading centers. Both of these models produce gaps in the seamount pattern if a single non-mobile "hot spot" is hypothesized.

(2). The spatio-temporal arrangement of the Easter Island seamounts does not agree in detail with the "hot spot" hypothesis (Harrison et al., 1974).

(3). The data may be more logically interpreted by hypothesizing: (a) that the hot spot "jumps" with the ridge crest, or (b) that there are two hot spots that move relative to each other.

(4). The zone of production may be much wider than is implied by the hot-spot concept, suggesting that the seamounts form over a broad belt with variations in age of the seamounts explained by variations within the broad belt.

(5). From this, it appears that Concept B is not supported by observational evidence regarding actual, real-world seamounts, and that Concept A provides a better approach to obtaining an understanding of the geometrical, thermal, mechanical, and chemical aspects of the development of the surficial manifestations of the Earth's internal thermal behavior.

A further discussion of the hot-spot concept is not essential to this analysis of the surficial manifestations of the Earth's internal thermal behavior. However, many discussions exist in the literature for further reference (See, e. g., Vogt, 1974, and the references cited therein).

Intrusive and Extrusive Activity.

In considering the surficial manifestations of the Earth's internal behavior from the standpoint of intrusive and extrusive activity on a global scale, it is necessary to consider all types of intrusive and extrusive activity that appear to represent mantle material or which appear to be produced by deep-seated tectonics. This includes volcanic activity, the creation of basaltic traps both in continental and oceanic areas, mantle upwelling at crustal rifts, the emplacement of batholiths, laccoliths, and other intrusive bodies during at least the past 3.6 billion years.

What, if any, global patterns are formed by the Earth's intrusive and extrusive activity? To answer this question, it would be well to consider the following observations regarding the Earth's intrusive and extrusive activity: (a) it is usually agreed that the ultramafic rocks contained in layered, stratiform, and other intrusions involving gabbro and dibase, together with accumulations in concentrations of mafic material, were formed from mantle-derived basaltic magma (See, e. g., Wyllie, 1969); (b) all other ultramafic rocks appear to be associated with major tectonic features of the Earth's crust with distributions controlled by deep-seated tectonics with linear trends; (c) Alpinotype ultramafic rocks are presently found along deformed mountain chains and island arcs, usually with associated gabbro or basic volcanic rocks; (d) the occurrence of serpentinites and peridotites along mid-ocean ridges suggests a third type of ultramafic belt, but exactly how it may be related to the other two is not known; (e) although the petrogenesis is extremely complicated in some cases by possible "coupling" between mantle-derived and crustal material, there are reasons to believe that all ultramafic rocks are derivable ultimately from a mantle source; (f) most observers feel that the petrogenesis of Alpine intrusions in orogenic regions is complicated by the fact that the petrogenesis normally involves several destructive metamorphic episodes but that Alpine intrusions represent parts of a solid, or partially-fused, mantle that has flowed into or through the crust along an unstable, linear orogenic belt; (g) vast composite sheets of Mediterranean and Himalayan areas appear to represent voluminous extrusions of mantle material breaking through the floor of the Tethyan sea-bottom to depths of 8 to 10 km (See, e. g., Maxwell, 1968); (h) genetic links between oceanic ultramafic rocks and Alpine ultramafic belts are suggested by the observation that Puerto Rican serpentinites represent uplifted oceanic crust composed of altered mantle material exposed at the surface, as well as by the observation that a "spreading" ocean floor could cause tectonic incorporation of the serpentinites of the oceanic crust into the underlying sediments of the continental rise, thus producing ultramafic rocks of the Alpine type upon metamorphism of the sedimentary pile (See, e. g., Wyllie, 1969); (i) similarities between the basic pillow lavas and peridotites which appear to characterize the mid-ocean ridges (Maxwell, 1968) and the extrusions of mantle material in the ophiolite complexes of the Alpine mountain belts (i. e., those formed during the past 200 m. y.) suggest that mid-ocean ridges may be the loci of ultamafic belts just as extensive as the Alpine ultramafic belts within land areas (See, e. g., Wyllie, 1969).

From the above and related analyses (See, e. g., Chao, 1973; Drake, 1969, 1970; Drake and Kosminskaya, 1969; Green, 1969; Gurbanov, 1972; Harrison, 1972; Kisvarsanyi, 1973; Lanphere and Reed, 1973; Mannard, 1973; Pereira and Dixon, 1971; Petraschek, 1973; Richards, 1971; Russell, 1973; Sawkins, 1972; Sillitoe, 1972; Simmons, 1969; Stanton, 1972; Tarling, 1973; Vogt, 1974), it may be seen that, when viewed on a global scale, intrusive and extrusive activity, of both the mid-ocean-ridge type and the Alpine-ultramafic-belt type, may be considered to be surficial manifestations of a single, global mechanism characterized by cyclical-but-aperiodic activity along long, linear belts of deep-seated control. Closer scrutiny reveals that intrusive and extrusive activity forms a complete spatio-temporal spectrum, or continuum, within each cycle of orogenic activity whether of the mid-ocean-ridge type of of the Alpine-ultramafic-belt type.

Mountains and Geosynclines.

The relationship that exists between mountains and geosynclines is not completely understood. It appears, however, that there may be an antithetical, Phoenix-like association between them and that the general nature of this association might become more obvious if mountains and geosynclines are analyzed simultaneously as surficial manifestations of the Earth's internal behavior on a global scale during the 4.6 b. y. that the Earth is believed to have been in existence. The most obvious feature common to both mountains and geosynclines is their linearity when viewed on a global scale. Also, both appear to go far back into the history of the Earth, perhaps back at least 3.6 b. y. Following are other significantly interesting data regarding mountains and geosynclines: (a) a geosyncline appears to be an elongate downwarp in the Earth's crust, the bottom apparently subsiding deeply beneath accumulating sediments; (b) the Allegheny mountain range evolved from a long-continued linear subsidence, followed by an accumulation of sediments during the entire Paleozoic (a span of 300 m. y.), completed by great breakings, faultings, and foldings of strata along with other results of deep-seated disturbances; (c) since it appears that the deposition of sediments was in shallow water, some observers (See, e. g., Hall, 1857) feel that the geosynclinal depression was caused by the weight of the accumulating sediments, but others (See, e. g., Dana, 1873) see evidence suggesting that the accumulation of sediments was a consequence, rather than a cause, of the subsidence; (d) some, but not all, mountains and geosynclines appear to form along edges of continents; some appear to form entirely within continental areas; some seem to form partly in and partly out; and, if crustal rifting is to be considered as the earliest stage of the mountain-building cycle, then some may form completely within oceanic areas; (e) particularly when the paired-belt concept of miogeosynclines (nonvolcanic) and eugeosynclines (volcanic) is used (See, e. g., Stille, 1940), the identification of geosynclines along continental margins becomes confusing to the circumspect analyst and often appears meaningless because the spatial separation between the mio and eu elements of the geosynclinal pair is observed to be more than a thousand km in many case; (f) there are indications that many eugeosynclines (e. g., within the American Cordilleran and Appalachian systems) were analogous to present island arcs and that sediments were derived from active volcanoes within the sedimentary basins associated with them (See, e. g., Kay, 1951); (g) there is evidence that both present and past geosynclines formed in an orderly manner in all parts of the globe, suggesting the possibility of orderly control on a global scale (See, e. g., Drake et al., 1959; Dietz, 1963; Dietz and Holden, 1963; Tatsch, 1972a); (h) there is evidence that much of the "fill" for the North American geosynclines was derived from the "borderlands" situated on the oceanic side of the continental margins (See, e. g., Hsu, 1965); (i) the structure and development of the Alpine geosyncline seem to have been similar to those of the Indonesian arcs, consisting of a complex system of troughs and ridges throughout most of its history (See, e. g., Aubouin, 1965); (j) when geosynclines are considered to form along all the continental margins, there are five types of geosynclines: Atlantic, Andean, island-arc, Japan-Sea, and Mediterranean (Mitchell and Reading, 1969); (k) on a global scale, these five types of geosynclines may be regrouped into two basic types (Tatsch, 1972a): those occurring near a "mobile" belt (the Andean, island-arc, Japan-Sea, and Mediterranean types) and those not occurring near a mobile belt (the Atlantic type); (l) in the first of the regrouped types, there is both associated orogeny and differential plate movement, whereas, in the second, there is neither.

From the above and other analyses (See, e. g., Argand, 1916; Baer, 1970; Bonini et al., 1973; Borrello, 1972; Burtman, 1972; Cepeda, 1973; Dickinson, 1973b; Dott, 1972; Ernst, 1973a, 1973b; Fisher et al., 1970; Gilluly, 1971; Graham et al., 1974; Haile, 1973; Hoffman, 1973; Hawkesworth et al., 1975; Kanasewich et al., 1973; Kuenen, 1967; Laubscher, 1969; Marlow et al., 1973; Pautot et al., 1973; Peive et al., 1972; Pushcharovsky and Arkhipov, 1972; Schneider, 1972; Scholl and Marlow, 1974; Schuchert, 1923; Shtreys and Tseysler, 1972; Silver et al., 1974; Stewart, 1972; Sylvester-Bradley, 1968; Zapletal, 1968; Zonenshain, 1972), it appears that, when geosynclines and mountains are analyzed simultaneously on a global scale and as "active" features (i. e., excluding the "fossil" or inactive Atlantic-type geosynclines), then their origin and evolution appear to be controlled by a single, global mechanism deep within the Earth (Tatsch, 1972a: chapter 12).

The Earth's Gravity Field and the Shape of the Geoid.

If the Earth's gravity field and the shape of the geoid are surficial manifestations of the internal behavior of the Earth during the past 4.6 b. y., then they should have certain characteristics that are correlatable with the Earth's internal behavior on a gloaal scale during that period of time. Under ideal conditions of isostatic equilibration, all crustal columns exert equal pressure at some depth, and the column extending above sea level has a mass exactly equal to that of the compensating mass at depth. Much has been written in this area (See, e. g., Walcott, 1972; Artyushkov, 1974; Heiskanen and Vening Meinesz, 1958; Heiskanen and Moritz, 1967; Caputo, 1967; Woollard, 1969a, 1969b, 1969c, 1972; Kaula, 1972). Although the Earth does not behave as an ideal isostatic system, much can be gained from a brief analysis of the nature and extent to which the Earth's internal behavior departs from that of an ideal isostatic system when viewed on a global scale, over a long period of time.

Particularly germane to an analysis of the Earth's surficial manifestations in this connection are the following observations related to the isostatic behavior of the Earth: (a) significant changes (25 to 50 mgal)in isostatic anomalies are to be expected where there are changes in crustal composition (horizontal and vertical heterogeneities); (b) there is a strong correlation among gravity values and topographic relief, changes in geology, and areas of change in crustal composition and thickness, particularly when these areas are narrow; (c) there appears to be a slight elevation dependence, governed more by the wavelength (width) of a particular topographic feature than by its amplitude (elevation); (d) major changes appear to result from variations in crustal and upper mantle parameters, including deep mass anomalies; (e) in cases of major changes, sea-level intercept values may vary by as much as 50 mgal; and (f) generally speaking, gravity-anomaly gradients exceed normal values in the case of graben-type surficial features and are subnormal in the case of horst-type surficial features.

Thus, when viewed on a global scale, it appears that the Earth's gravity field and the shape of the geoid may reflect the Earth's internal behavior during the past 4.6 b. y., provided that appropriate spatial and temporal corrections are made for the fact that the Earth is a dynamic body and that its surficial features may lag behind its deep internal behavior. For example, corrections must be made for the fact that ideal conditions of isostatic equilibrium exist if, and only if, the unbalanced force acting on a vertical column is zero; in all other cases, the column tends to move in a direction parallel to that of the unbalanced force vector. It is extremely difficult to visualize a case of absolute ideal isostatic equilibrium within the real dynamic Earth, and actual gravity surveys appear to confirm this. However, approximate isostatic balance of subglobal or regional features would be expected, and such is observed, but not in agreement with present continental and oceanic positions. The independence of geoidal and continental features compels the inference that geoidal features are due to either (1) density differences deep within the mantle (i. e., deeper than the so-called low-velocity layer in the upper mantle, by virtue of which isostatic balance might be feasible); or (2) density differences maintained by convection, or by a similar mechanism, in which case they are more apt to lie along the upper mantle. There is no reason why they connot be a combination of both (1) and (2), that is, with rheid surfaces, shells, and/or planes at more than one depth within the tectonosphere (See, e. g., Tatsch, 1972a: chapter 6).

On the basis of the present evidence, it appears that the Earth's gravity field and the shape of the geoid are surficial manifestations of the internal behavior of the Earth during the past 4.6 b. y. However, it appears also that mass redistributions lag behind the forces causing the Earth's internal behavior. The shape of the geoid, therefore, should correlate with the tectonospheric driving mechanism rather than with the surficial manifestations of the Earth's internal behavior (Tatsch, 1972a).

Geomagnetism and Polarity Reversals.

If geomagnetism and polarity reversals are surficial manifestations of the Earth's internal behavior, then the following should be applicable in connection therewith: (1) the geomagnetic dynamo has evolved and is driven by the same causal mechanism that has produced the Earth's internal behavior during the past 4.6 b. y.; (2) the geomagnetic field correlates, now and in the past, with global characteristics of the Earth's driving mechanism; and (3) cyclical-but-aperiodic polarity reversals are definable in terms of the geometry and mechanics of the Earth's driving mechanism.

Most geomagnetic dynamos fall into one of two categories: (1) the convective geomagnetic dynamos which derive their driving force from convective processes deep within the Earth; and (2) precessional dynamos which derive their driving force from the Earth's precession.

Theoretically, a convective dynamo could have evolved from an adiabatic temperature gradient in the outer core, resulting from the latent heat of a progressively solidifying inner core

(See, e. g., Verhoogen, 1969). However, such a mechanism would not have maintained a convective dynamo for more than about 3 b. y. For this and other reasons, a precessional dynamo might be preferred to a convective one (See, e. g., Malkus, 1963). There is a possibility that the Earth's actual dynamo is neither convective nor precessional (Tatsch, 1972a: chapter 14).

In order that we may determine the true nature of the geomagnetic dynamo, it is necessary to examine the Earth's thermal behavior during the past 4.6 b. y. For example, if the Earth's core is in a state of thermal convection, then it is cooling at a rate of about 100° C per b. y. In the case of the precessional dynamo, nothing need be said about deep cooling, although lunar precession implies a decrease in the internal heat generated by the precessional torques. Thus, the heat flux from the Earth's core need not be appreciably different for the two basic types of dynamo because a "stirred" outer core implies an adiabatic temperature gradient against conduction in both types.

Present-day observations show that the dipole moment of the Earth is changing by about 1/20 of 1% each year, indicating a time constant of only about 2000 years, from which it follows that the geomagnetic field may not be simply a decaying vestige of the primordial Earth but must be continuously motivated by some means or other (Tatsch, 1972a: chapter 14).

The motive force for the geomagnetic dynamo poses no problem if the Earth's rotation enters into the system (See, e. g., Bullard, 1949). For example, since the core is considered to be ellipsoidal rather than purely spherical, it may not follow mantle precession exactly, with the result that internal motions must occur and that they might be more adequate to drive a geomagnetic dynamo. However, the issue is still in doubt because the dynamo currents appear to be thermo-electric in origin rather than purely thermal. Also, the dominance of the dipole field and its axial character (at least over periods of tens of thousands of years) suggest that the Earth's rotation (or some other axi-symmetric characteristic) exercises a controlling influence, leading to the conclusion that the motive power for the geo-magnetic dynamo might be a linear combination of several forces.

Geomagnetic polarity reversals appear to occur with greater frequency during periods of rapid polar wander and/or continental drift (See, e. g., Irving, 1966). This is a particularly intri-guing observation, because it suggests that the configuration of the mantle might have a direct influence upon the core motions which are thought to be responsible for the geomagnetic field. In an attempt to resolve this paradox, several conclusions may be drawn. First, it is difficult to explain the correlation between the frequency of polarity reversals and the rate of polar wander and/or continental drift on the basis of ordinary convec-tive dynamo processes. Rather, this correlation appears to be somewhat relatable to the precessional dynamo, because the process of polar wander would appear to cause the axis of symmetry of the core to depart somewhat from the axis of rotation of the Earth and, thereby, to alter the differential precessional torques acting on the core and mantle. What seems completely unresolved, however, is that the geomagnetic dynamo with a computed time constant of only about 10^3 years could be sensitive to axial changes on a time scale of 10^6 years, or 3 orders of magnitude greater.

Electrical conduction in the lower mantle, such as would be required in a thermo-electrical dynamo, may result from the semiconducting properties of silicates at elevated temperatures. Other observations of interest include the following: (1) the average crust is less conductive than are sedimentary rocks; (2) conductivity is increased by the amount and the salinity of interstitial water; (3) the mantle shows a rapid increase in conductivity with depth; (4) the upper-mantle conductivities may vary by two orders of magnitude within small horizontal distances (e. g., the upper mantle conductivities under Arizona are 50 to 100 times greater than are those under New Mexico); and (5) conductivities are higher in regions of high heat flow rates and low seismic velocities (See, e. g., Madden and Swift, 1969).

It is generally concluded that the nondipole components of the geomagnetic field reflect some characteristics of deep-seated features, but the details are not clearly understood (See, e. g., Stacey, 1969; Doell and Cox, 1972). This arises partly from the fact that spherical harmonic analyses of the geomagnetic field do not produce individual harmonic terms identifiable with features of the field except perhaps for the dipole terms. Other analyses suggest that the field may be due to several deep-seated dipoles (See, e. g., Bullard, 1956). It appears that eight non-central radial dipoles reasonably represent the field when considered on a global basis, with the dipoles at about ¼ Earth radius, or possibly farther, from the center (See, e. g., Lowes and Runcorn, 1951; Alldredge and Hurwitz, 1964).

In any analysis of the geomagnetic polarity reversals, the above observations about the geomagnetic and geoelectric fields must be supplemented by considerations regarding the applicable global frames of reference, as well as the geometrical and mechanical environment in which the fields exist (Tatsch, 1972a: chapter 14). If one had measureable inducing fields with a broad spectrum of frequencies and length scales to probe the entire Earth (or at least certain selected portions of it), then it might be possible to achieve a satisfactory solution to the problem of the actual distribution pattern of electrical conductivity within the mantle. Unfortunately, such is not practical, and it is possible to observe only the natural magnetic variations, that might be considered to reflect the geoelectric field. However, these variations are usable at only relatively few, discrete frequencies, in a narrow bandwidth, with the result that the conductivity solution is subject to large systematic errors. The situation is further aggravated by the following unknowns: (1) the effects of the oceans and sedimentary regions and pockets; (2) the lack of knowledge regarding the size and nature of the so-called "low-conductivity layer"; and (3) the role of heat flow.

Symmetry and other geometrical considerations seem to enter into the origin, evolution, and present characteristics of the geomagnetic field. For example, the observation that the geomagnetic field is roughly symmetrical about the Earth's axis of rotation has at least two important implications: (1) it has an important bearing on paleomagnetic analyses, and (2) it poses important constraints on hypotheses regarding the origin and perpetuation of the Earth's magnetic field (See, e. g., Stacey, 1969).

Within the geological time scale, the angular variation (inclination) between the geomagnetic axis and the Earth's rotational axis appears to be a transient. Consequently, there is little reason to search for asymmetric elements in the Earth or in geomagnetic dynamo mechanisms hypothesized to explain the evolution and present behavior of the geomagnetic field. Although the principles of symmetry are normally applied intuitively to physical cause-and-effect situations, a more explicit approach should be used in cases such as this where cause-and-effect relations are closely intermingled (and even, on occasion, possibly interchanged) (See, e. g., Elsasser, 1966).

As far as can be determined, the relevant contributing causes of the geomagnetic field are simply the rotation of the Earth (which, for all practical purposes, is axisymmetric) and various motions, temperature gradients, etc., in the core, all of which, apart from the effect of rotation, are presumed by most observers to be spherisymmetric. Thus, if the geomagnetic field were to deviate in a consistent manner (asymmetric) with respect to the axis of rotation, then the situation would be that of a field with lower symmetry than that attributable to a combination of the known contributing causes. This would require us to search for some other (currently unknown) contributing cause of lower symmetry. However, since, there is no consistent inclination of the dipole field, there is no need for such search. Thus, it is reasonable to assume that the geomagnetic axis is constrained to undergo transient variations about the Earth's rotational axis as an axis of symmetry; but the nature of the causal mechanism for such constrained control is not known.

Evidence for the reality of geomagnetic polarity reversals on a global scale over long periods of time may be summarized: (1) there is a marked correlation between polarity reversals observed on different continents and within different rocks on a single continent in the case of both igneous and sedimentary complexes; (2) sediments which were baked by contact with subsequently emplaced igneous rocks have acquired thermoremanent magnetism of the same polarity as the igneous rocks in almost all cases; and (3) the actual process of reversals has been traced both in rapid sequence of lava flows and in deep-sea cores (See, e. g., Stacey, 1969).

Fluctuations in the polarity of the geomagnetic field are known only for the past 4.5 m. y. (i. e., during the past one-tenth of one percent of the Earth's existence) but enough events are involved to permit certain conclusions to be drawn on the basis of what has happened during this relatively short period: (1) reversals appear to have occurred cyclically, but in an aperiodic manner, on an average of once every 100,000 yr.; (2) each reversal process lasts about 5000 yr (i. e., about 5% of the cycle); (3) the magnitude of the global field strength is reduced by a factor of 4 or 5 during the reversal process, suggesting that the nondipole field and possibly an equatorial component of the dipole field may remain effective during the actual reversal process; (4) tht total number of magnetized rocks is about equally divided between normal and reversed polarities; (5) the reversals are complete 180° reversals (rather than a succession of increments totaling 180°); (6) the basic causes contributing to the geomagnetic field appear to be axisymmetric without polar symmetry, i. e., there is no distinction between the opposite axial directions, either polarity being equally probable; (7) during certain periods, such as the Permian, the geomagnetic field apparently had a constant polarity (i. e., without detectable reversals) for tens of millions of years; and (8) reversals appear to have occurred more frequently during periods of rapid polar wander and/or continental drift, thereby implying that mantle configuration may have a direct influence upon those core motions that are responsible for the geomagnetic field (Irving, 1966).

An analysis of the above indicates that the following tentative conclusions may be drawn: (1) that the geomagnetic dynamo has evolved, and is presently driven, by the same global causal mechanism that has produced the Earth's internal behavior during the past 4.6 b. y.; (2) that the geomagnetic field correlates, now and in the past, with global characteristics of the Earth's driving mechanism; and (3) that the cyclical-but-aperiodic polarity reversals are definable in terms of the geometry and mechanics of the Earth's driving mechanism. On this basis, it appears that, when viewed on a global scale, geomagnetism and polarity reversals are surficial manifestations of the Earth's internal behavior and that such has been the case for at least part of the past 4.6 b. y. (Tatsch, 1972a: chapter 14).

Continents and Oceans.

In order to determine whether continents and oceans are surficial manifestations of the Earth's internal behavior, it is necessary to (1) analyze the interactions now occurring between continents and the floors of the oceans, and (2) to consider types of interaction that might have occurred during the past 4.6 b. y., with particular attention to global and regional interactions that might be interpreted as having deep-seated causal connections.

It is generally accepted that there are two basic types of continental margin differing quite significantly from each other. From geographical associations, representative members of these are known as the Atlantic and the Pacific type margins (See, e. g., Drake, 1969).

The Atlantic type continental margin has stable unmodified blocks on the landward side and no evidence of post-Paleozoic activity. Ocean-to-continental crustal dimensions in Atlantic type margins are rather abrupt and may, in places, be characterized by deep sedimentation. Atlantic type margins do not normally contain deep trenches, although trench-like features may be found beneath the sediments of the associated continental rise. In many cases, Paleozoic and earlier features are truncated by the continental margin.

The Pacific type continental margin, on the other hand, has young tectonic belts parallelling the shores. Shallow-focus earthquakes are very common, and intermediate and deep-focus earthquakes are abundant, particularly along certain segments of the Pacific type margins. Volcanic activity occurs, or has occurred historically, along the entire Pacific belt. Many portions of the margins contain deep trenches. Ocean-to-continental transition is much more variable than in the Atlantic type, and sedimentary accumulations do not appear to be great.

Analyzed on a global scale, it appears that the difference between Atlantic and Pacific type margins may be a function of position with respect to an active portion of the Earth's driving mechanism, as previously discussed in connection with geosynclines (page 51 et seq.).

A deeper analysis leads to the question as to whether the above categories of margin may be applicable to the oceans and the continents themselves, as well as to their margins. To answer this question, it is necessary to analyze not only the oceans and continents as surficial manifestations of the Earth's internal behavior but also with respect to the understructures beneath the oceans and continents. Also, it is essential that such be done over a long period of time, because it is reasonably well established that the continents and oceans have not always been where they are today.

As long as the continents are exposed, erosion tends to lower sea level, which action disturbs the isostatic balance and renews uplift; but, if the sea level is raised, then the flooding of the continents produces no such restoring effect. The elevation of most continents averages less than a few hundred meters above sea level; the most-elevated continent (Eurasia) averages only about one km. Several processes tend to raise sea level, including deposition of the products of erosion in the oceans, the release of new sea water, and the addition of fresh lava. The effect of these processes is to reduce the area of the ocean basins and to deepen the ocean. Such being the case, it is surprising that most of the continents were not long ago submerged by shallow seas.

The above suggests that there must be long-lived, deep-seated global forces tending to increase the heights of the continents (or, more generally stated, to increase the differences in the "heights" of the continents and oceans). Such forces are not in evidence on the surface and, being imperceptible, are the subject of speculation. This is particularly true regarding the cyclical-but-aperiodic uplift of continental areas, which seems to have taken place on three scales, i. e., involving areas having diameters of thousands, hundreds, and tens of km, or perhaps a complete continuum of diameters ranging from km to thousands of km.

For our purpose, the important point is that most of the presently-observable uplifts are Tertiary in age and that they represent uplift of land over very extensive areas. Some continents (e. g., Africa and Antarctica) have been subjected to recent uplift; others (e. g., Australia) have not. In other regions of the globe, only portions of continents have undergone uplift, e. g., the Colorado Plateau and the Basin and Range province of North America.

Alternating with uplift, there have been periods of subsidence and marine invasion. For example, the central portion of North America has been widely invaded by waters, roughly once every 100 m. y., on average, since at least the Precambrian, alternating with periods of uplift and withdrawal of the sea. These sequences are temporally independent of similar activity in other parts of the Earth. In fact, these are temporally distinct from those of the Appalachian and Cordilleran belts on the same continent.

Other geometrical, mechanical, thermal, and chemical aspects of continents and oceans have been investigated by Beloussov (1971, 1972), Bott (1973a, 1973b), Coleman (1973), Flemming and Roberts (1973), Garrels et al. (1972), Hailwood and Tarling (1973), Heacock (1971), Heirtzler (1973), Jackson et al. (1972), Johnson (1973), Laughton et al. (1973), Mitchell-Thome (1972), Mouratov (1972), Ostenso (1973), Reyment (1973), Ross and Schlee (1973), Sclater and McKenzie (1973), Sclater et al. (1975), and Zijderweld and Van der Voo (1973).

Viewed on a global scale, it appears that most areas, whether oceanic or continental, have suffered as many as 100 episodes of uplift during the past 3.6 b. y. Consequently, present continental areas have not necessarily always been continental nor have present oceanic areas necessarily always been oceanic. From this, it appears that oceans and continents may be surficial manifestations of the Earth's internal behavior to the extent that the Earth's driving mechanism appears to have given the Earth's surface an ever-changing temporal and spatial panorama of continents and oceans in all parts of the globe during at least the past 3.6 b. y. (Tatsch, 1972a: chapter 15).

Crustal Rifting and Sea-Floor Spreading.

In order to determine whether crustal rifting and sea-floor spreading are surficial manifestations of the Earth's internal behavior, it is necessary to understand the exact nature and possible inter-relationships of these two phenomena. Basically, when viewed on a global scale and over a long period of time, it appears that crustal rifting occurs on certain parts of the globe when surficial manifestations of the internal behavior appear in the form of crustal "tension". Since crustal rifting of this type was originally thought to occur only in the oceans, it was called "sea-floor spreading". Present evidence shows that oceanic crustal rifting sometimes extends into continental areas (e. g., North America, Central America, Iceland, and Africa). In spite of this, the original term sea-floor spreading is often used even when referring to the type of crustal rifting that extends into continental areas. Such being the case, this book will use the term "seafloor spreading" in those cases where current usage dictates. However, the author prefers the more-general term, "crustal rifting", for both oceanic and continental rifting of this type.

According to the sea-floor spreading concept, new oceanic lithosphere is formed more-or-less continuously at certain mid-ocean ridges and then spreads outward at rates of a few cm/yr (See, e. g., Burke and Whiteman, 1973; Deutsch and Murthy, 1973; Francheteau, 1973; Francheteau et al., 1972; Heirtzler et al., 1968; Le Pichon, 1968, 1969; Lowell and Genik, 1972; Pittman and Talwani, 1972; Vine and Matthews, 1963; and Wright and McCurry, 1973). Many geological observations indicate that sea-floor spreading, as defined above, has been occurring for at least the past 100 m. y. This is based on evidence that most ocean floors are no older than that. Because continents are at least an order of magnitude older than are the oceans, this suggests that a process similar to the above might have created the continents during the past 3.6 b. y., while repeatedly renewing the oceans.

Wilson (1965) reported evidence that the ages of fossil volcanoes (seamounts) increase with their distances from the ridges, suggesting that these fossil volcanoes were once active volcanoes near the ridges. However, in deriving the locations of fossil volcanoes from those of active volcanoes, it is well to realize that sea-floor spreading probably does not occur at a linear rate nor at the same rate throughout the entire surface of the Earth at any given time. Most evidence suggests a cyclical-but-aperiodic rate for this spreading, which changes direction as the unbalanced force vector changes (Tatsch, 1972a).

Because sea-floor spreading is commonly attributed to some type of mantle convection as a "driving force", Lee and MacDonald (1963) attempted to obtain evidence of a convective pattern by comparing spherical harmonic analyses of gravitational potential and surface heat flow, on the assumption that a correlation between geoidal lows and heat-flow highs would represent rising limbs in the hypothesized convection pattern. Their results were not conclusive, primarily because the amount of data available was too small for the particular type of correlation attempted. Subsequent similar analyses have not clarified the nature of the suspected relationship between sea-floor spreading and mantle convection, other than to suggest that both sea-floor spreading and the hypothesized mantle convection might be surficial manifestations of the Earth's internal behavior.

Summarizing, it may be said that crustal rifting (and sea-floor spreading as a subsidiary phenomenon thereof) appears to be a surficial manifestation of the Earth's internal behavior. Furthermore, such crustal rifting appears to occur when and wherever the Earth's internal behavior causes crustal tension (Tatsch, 1972a).

"Plate Tectonics" and Related Matters.

If plate motions are a surficial manifestation of the Earth's internal behavior, then the behavior of the plates should be correlatable with certain characteristics of the Earth's internal behavior on a global scale and over a long period of time.

Basic to the plate-tectonics concept is the idea of a rigid upper layer of the Earth (lithosphere), composed of the crust and uppermost part of the mantle. The lithosphere is assumed to have considerable strength, to be roughly 100 km thick, and to rest or float upon a second layer, the asthenosphere, which is assumed to have practically no strength and to extend from the base of the lithosphere to a depth of several hundred km below. The floating lithosphere is divided into vast plates or blocks which are bounded by ocean ridges, by certain faults, and by the great system of quasi-arcuate structures such as the circum-Pacific system. The plates spread apart at the ridges, grind against each other at the faults, and are underthrust at the island arcs and similar structures (See, e. g., Drake, 1970).

The basic plate-tectonics concept hypothesizes six major plates (and some minor ones). The major ones are: (1) the American Plate, including North and South America and the western half of the Atlantic floor; (2) the African Plate, including Africa, the eastern half of the South Atlantic floor, and the Indian Ocean floor out to the Mid-Indian Ridge; (3) the Eurasian Plate, including Europe and Asia except for India and Arabia; (4) the Pacific Plate, containing both the North and South Pacific; (5) the Australian Plate, including Australia plus some less well-defined ocean areas; and (6) the Antarctic Plate, including Antarctica plus some less well-defined ocean areas.

To cover the entire globe, the plate-tectonics concept supplements the above six major plates by various minor plates, such as the Caribbean, Cocos, Nazca, Philippine, Indian, Arabian, and Aleutian plates. These plates, both major and minor, move and supposedly have been moving over the surface of the Earth during at least part of the past 3.6 b. y. They have been carrying with them the continents and/or oceanic sediments and have been spreading apart at the ocean ridges where hot basaltic material from the mantle rises and joins the trailing edge of the plates.

The plate-tectonics concept hypothesizes that major collisions between plates have created, over at least most of geologic time, some of the Earth's major topographical features. Minor collisions cause earthquakes. If the velocity of the collision is less than 6 cm/yr, then the plates tend to buckle and fold, creating ranges of new mountains, such as the Himalayas (See, e. g., Le Pichon, 1968). If the velocity of collision is more than 6 cm/yr, then one plate thrusts itself under the other and descends into the asthenosphere at an angle of about 45°. Upon reaching the mantle, it is hypothesized in this concept, the leading edge of the underthrusting plate is progressively reconverted into mantle material. The underthrusting plate creates the deep, linear trenches that are observable in locations such as Tonga, the Kuriles, Japan, the Marianas, the Peru-Chile coast, Puerto Rico, New Zealand, etc.

The plate-tectonics concept further hypothesizes that the impact of two colliding plates produces the islands and volcanoes, and that the impact accounts for major earthquakes observed in the vicinity of the trench-and-island-arc systems. Collisions at the edge of a continent produce long quasi-linear mountain ranges, such as the American Cordilleras, which stretch from Alaska to Tierra del Fuego, a distance of about 15,000 km, on the assumption that some of the ocean-bottom sediments on the underthrusting plate are scraped off and piled up along the edge of the under-thrust continents. The heat of the scraping supposedly causes volcanoes to develop in the piled-up sediments (See, e. g., Dewey and Bird, 1970; Isacks et al., 1968; Le Pichon, 1968; Mc-Kenzie and Parker, 1967; Morgan, 1968).

The origin and evolution of the forces that supposedly have driven these huge plates over the surface of the Earth, during at least part of the past 3.6 b. y., have not been defined in any detail by the plate-tectonics concept, Nor has the plate-tectonics concept defined the geometrical and mechanical frame-work which has controlled these forces during all this time (See, e. g., Burke and Dewey, 1973; Cady, 1973; Clague and Jarrard, 1973; Dewey and Burke, 1973; Dewey et al., 1973; Dickinson, 1972; Hamilton, 1973; Sipkin et al., 1975; Terman, 1974; Watson, 1975).

The basic plate-tectonics concept leaves several questions unanswered. What, for example, has initiated the global driving mechanism that has evolved, even perhaps syngenetically with the Earth, 4.6 b. y. ago, or at some time since then, say 3.6 b. y. ago? Once initiated, what has sustained this fairly well-organized driving mechanism during the past 4.6 or 3.6 b. y.? To answer these questions, consideration must be given to the development of the forces that appear to have been both long-lived and quasi-organized in a manner sufficient to perform this function in the internal behavior of the Earth and in the creation of its kaleidoscopic surficial features during at least part of the past 3.6 b. y.

One possible driving mechanism could have been the convection within the mantle created by differences in temperature under oceans and continents. Once started, the convection could then slowly move like a conveyor belt, splitting the plates apart to form a ridge and then carry them along on the huge conveyor-belt system. But, granted that forces were available to have started the system, say 4.6 or 3.6 b. y. ago, what would have kept it going? If it started, not 4.6 or 3.6 b. y. ago, but at some time since then, say 1 b. y. ago or 200 m. y. ago, what made it start then, rather than at some other time? Does this mean that, perhaps, the driving mechanism started syngenetically with the Earth, 4.6 b. y. ago, and that it has been going ever since?

Is there an alternative to the above system for splitting and moving the plates? For example, are gravitational forces, acting on the downthrusting lithosphere plate, exerting a pull sufficiant to drag it downward at an angle of about 45° to depths of at least 700 or 800 km? Or, perhaps, does the underthrusting plate cool the mantle to produce the convection currents that move the plates?

The geometry and mechanics involved in these and other questions are examined elsewhere (Tatsch, 1972a: chapter 18). Suffice it here to say that the plate-tectonic behavior appears to be a surficial manifestation of the Earth's internal behavior on a global scale during at least part of the past 3.6 b. y. However, the geometry and mechanics of the Earth beneath the plates, to a depth of at least 1000 km, must be more completely defined and analyzed than they have been thus far if we are to understand the Earth's plate-tectonic behavior.

Further analyses are also needed to explain surficial manifestations such as: (1) seismic activity underneath Nevada; (2) the creation of the Rocky Mountains and certain other mountain ranges that appear to be somewhat more than the plate scrapings predicted; (3) the existence of a complete spectrum of magma and associated rocks that appear to follow certain global distribution patterns, such as is suggested by the "Andesite Line" or by basaltic flows of voluminous proportions in many parts of the world. Answers are also needed for certain other related questions, including: (1) how the geomagnetic dynamo was initiated and maintained, over a period of at least several b. y., and why, if the plates are moving in a more-or-less random manner, the Earth has managed to reverse its polarity in a cyclical-but-aperiodic manner; (2) the nature and origin of the numerous rectilinear orogenic-tectonic features observed on the surface of the Earth and what mechanico-geometrical configuration causes them to be rectilinear, since Le Pichon's plates, by-and-large, do not appear to have rectilinear edges; and (3) why the leading edge of the downthrusting plate descends at an angle of about 45° (rather than vertically or at some other angle).

<u>Continental Drift and Polar Wandering</u>.

If portions of the Earth's surface are undergoing a certain amount of drifting and if, as a result or in addition to it, the Earth's pole of rotation is wandering, then these effects probably fall within the category of surficial manifestations of the Earth's internal behavior. What mechanism within the Earth's interior causes continental drift and polar wandering? Is this something relatively new, or has it been going on for the past 4.6 b. y.? If the latter, what has managed to move the continents over the surface of the Earth, during all these years? If this mechanism started recently, say 200 m. y. ago, what caused it to start then, rather than 300 m. y. ago, or 900 m. y. ago, or 2.9 b. y. ago? Is continental drift proceeding in a strictly random manner, or is it following a quasi-organized, or controlled, pattern? Is pre-drift continental splitting (e. g., Africa from South America) induced by random forces, or are they quasi-linear as would be expected if they were induced by an organized, non-random global driving mechanism operating over a long period of time?

The hypothesis of continental drift, which proposes that continent-sized blocks of the Earth's crust have retained their form but have moved relative to one another, is today fairly well established on the basis of global paleomagnetism analyses, supplemented by regional and global studies of other types. Although the earlier concepts of continental drift are associated with Du Toit, Taylor, and Wegener, the modern approach is usually associated with Runcorn (1962), Bullard (1964), Blackett (Blackett et al., 1965), Isacks (Isacks et al., 1968), and more-recent researchers.

General concepts and observations supporting continental drift and/or polar wandering may be summarized as follows: (1) the southern continents (Africa, Australia, South America, India, and Antarctica) appear to have been grouped around the latter during the Jurassic (say about 150 m. y. ago); (2) the polar-wander curves for Africa, South America, and Australia follow paths that are almost identical for the three continents and that have the shape of a large Z in which the diagonal is about 90°, or about 10,000 km, or a linear trace roughly one-quarter of the Earth's circumference; (3) polar-wander curves suggest continental-drift rates of 90° in 500 m. y., or about 2 cm/yr; (4) because other evidence suggests that large-scale tectonic processes occur at a few cm/yr, polar wander of this magnitude may be globally associated with at least some of the large-scale tectonic processes; (5) stresses supported by the lower mantle are too large to permit lower-mantle convection of a few cm/yr, equivalent to mantle-creep rates that are smaller than this; (6)both polar wandering and continental drift appear to be caused by the same general process of mass distribution as may be envisioned by a lower-mantle adjustment or equilibration of the equatorial bulge to a changing axis of rotation; (7) there could have been two separate super-continents (Laurasia and Gondwana) or only one (Pangea), but this concept may be too simple and temporally restrictive to the extent that it must be supplanted by one in which the super-continents have repeatedly broken up and rejoined in different global surficial configurations; (8) polar-wander and continental-drift episodes are not necessarily steady-state processes but appear to be cyclical-but-aperiodic processes in which relatively long quiescent periods are separated by short drift and/or wander episodes; (9) all polar movements relative to Australia, and extending over the 300-m. y. period from upper Silurian to mid-Cretaceous times, occurred within a 20 m. y. episode during the Carboniferous (about 300 m. y. ago) (Irving, 1966); (10) some correlation exists between drift episodes and the strong orogenic-tectonic activity that occurs within certain continents during episodes of drift and/or polar wandering; and (11) the ages of rocks are grouped in a manner suggesting a more-or-less cyclical sequence of tectonically active and quiescent periods, with the cycles occurring every 250 to 500 m. y.

Present astronomical determinations of longitude are too imprecise to indicate relative movements of a few cm/yr; however, latitude measurements made with reliable precision indicate a currently steady movement of the pole toward about 70°W at a rate of 10 cm/yr (Markowitz et al., 1964; Markowitz and Guinot, 1968). Francheteau and associates (Francheteau et al., 1969) have found evidence suggesting that the northwestern and northeastern Pacific have not always been rigidly connected to each other. From this, they concluded that the present Pacific plate (Le Pichon, 1968) is really a composite of at least two distinct plates and that volcanic, aseismic ridges within the Pacific Ocean may be the fossil boundaries of these plates now coalesced into a single plate. By plotting available geochronological data and zones of isochronous tectonic and igneous activity on a pre-drift reconstruction of Laurasia, Hurley (1969) has shown that the configurations of the age provinces and the prevailing trends of orogenic zones suggest that there were successively younger belts of tectonism around ancient super-continents. After a review of the observational evidence bearing on continental drift and polar wandering, Lyustikh (1969) concluded that present simple models must be extended and specialized to more-complex, realistic models that take into account real-world observations regarding the behavior of the Earth, its structure, and the inhomogeneities and anomalies of its interior, with particualr attention to the upper 1000 km, i. e., the tectonosphere.

From the above and other analyses (See, e. g., Ahmad, 1969; Allan et al., 1967; Anderle and Smith, 1969; Anderson, 1967, 1969; Athvale, 1973; Barazangi and Dorman, 1969; Beloussov, 1969a, 1969b; Brock, 1973; Brune and Dorman, 1963; Bullard et al., 1965; Colbert, 1973; Cox et al., 1964; Crawford and Wilson, 1969; Dietz et al., 1969; Creer, 1973; Dalziel et al., 1973; Atwater, 1973; Donaldson et al., 1973; Douglas et al., 1973; Ewing and Ewing, 1967; Fairbridge, 1973; Faure, 1973; Goldreich and Toomre, 1969; Gough, 1970; Griggs, 1939; Gromme et al., 1967; Hallam, 1973; Hein, 1973; Henderson, 1973; Hurley et al., 1967; Isacks et al., 1968; Johnson, 1967; Kanes et al., 1973; Keast, 1973; King, 1973; Knopoff, 1964, 1967a, 1967b, 1969; Knopoff and Shapiro, 1969; Le Pichon, 1968; Larson and Chase, 1973; Maxwell, 1969; McElhinny, 1973; McKenzie and Parker, 1967; Melville, 1973; Meservey, 1969; Morgan, 1968; Oliver and Isacks, 1967; Oliver et al., 1969;

Robinson, 1973; Romer, 1973; Runcorn, 1973; Salop and Scheinmann, 1969; Saxena, 1972; Smith, 1972; Spall, 1972; Tozer, 1965; Verhoogen, 1965; Vine, 1966, 1969a, 1969b; Vine and Matthews, 1963; Volker, 1973; Wegener, 1915; and Wilson, 1965), it may be concluded that continental drift and polar wandering are surficial manifestations of the Earth's internal behavior, during possibly the entire existence of the Earth, but that the geometry and mechanics of the Earth's interior and the nature of its driving mechanism remain to be defined.

The Earth's Most Probable Surficial Features 4.6 b. y. Ago.

In analyzing the present surficial features of the Earth that might have been caused by possible internal-behavior patterns of the Earth during the past 4.6 b. y., it is well to consider briefly the possible surficial configuration of the Earth 4.6 b. y. ago and at various critical times since then.

The configuration of the primordial Earth's most probable surficial features depends upon exactly which hypotheses are adopted to explore the evolution of the Earth from the solar nebula (Tatsch, 1972a: chapter 1) and to explain the evolution of the Earth-Moon system. If it is assumed that the Earth accreted from the solar nebula and that the Moon formed in the vicinity of (but not part of) the Earth, then the Earth's surface would probably have been devoid of any features other than those that would be genetic to an accretionary process. Thus, the Earth's surface would have presented a more-or-less homogeneous aspect 4.6 b. y. ago, both from the standpoint of topography (i. e., smooth) and from the standpoint of composition (i. e., the same) over the entire surface.

If, on the other hand, it is assumed that the Earth's surficial composition and/or topography were not homogeneous 4.6 b. y. ago, then such features would most probably not have survived to this date without some internal or external rejuvenation at least on one occasion since then. For the purpose of this analysis, therefore, it may be assumed that the Earth's surface 4.6 b. y. was devoid of any features other than those that would have been genetic to an accretionary process of planetary formation from the solar nebula at that time.

The Earth's Most Probable Surficial Features 3.6 b. y. Ago.

If it is assumed that the Earth had no appreciable surficial features 4.6 b. y. ago, when were the surficial features first acquired and what caused them to form? Were they acquired through external or internal behavior? If external, what was it? If internal, why did it occur when it did, rather than sometime earlier or later?

If the surficial features resulted from external sources during the period 4.6 b. y. to 3.6 b. y. ago, then such features would not now be active but would resemble the Precambrian shields and would have an age of about 3.6 b. y.

If the Earth acquired its surficial features from internal sources during the period 4.6 b. y. to 3.6 b. y. ago, then such features would not now be active unless the same causative internal behavior has been operating since 3.6 b. y. ago, in which case it would be reasonable to assume that the internal mechanism has been opertaing more-or-less continuously during the past 4.6 b. y. (because nothing internal to the Earth would have been available to activate it 3.6 b. y. ago).

In any case, it appears that the Precambrian shields existed 3.6 b. y. ago, but not necessarily in their present forms and locations, and that they have existed in one or more locations and forms, without having been severely heated or shocked, since then.

The Evolution of the Earth's Present Surficial Features.

Regardless of whether it is assumed that the Earth's surficial features of 3.6 b. y. ago were externally or internally induced, we must consider certain questions about what has happened to the Earth's surface since then. Have there been significant externally-induced catastrophes that might have altered the Earth's surficial features during the past 3.6 b. y.? If so, what were they, when did they occur, and what caused them? If no such external influences have operated during the past 3.6 b. y., is it reasonable to assume that all modifications to the Earth's surface since then were induced by the Earth's internal behavior?

To answer these and related questions about the evolution of the Earth's present surficial features, global-scale analyses must be made regarding both the present and past internal behavior of the Earth. Several international scientific bodies have been devoted to such analyses. In discussing the results of the International Upper Mantle Commission and the purpose of the Interunion Commission on Geodynamics, Drake (1970, 1973) has summarized many of the problems involved in analyzing the Earth's internal behavior and the surficial manifestations thereof. For the interested reader, Drake's papers constitute excellent summaries of the results of the International Upper Mantle Commission and the aspirations of the Interunion Commission on Geodynamics.

Other authors have made similar analyses but from different viewpoints or in less inclusive regions or aspects (See, e. g., Anderson et al., 1972; Clark et al., 1972; Cloud, 1972; Delany and Smith, 1969; Hart, 1969; Knopoff et al., 1968; Press, 1972; Stacey, 1969; Takeuchi et al., 1967; Wynne-Edwards, 1969). Earlier analyses of the Earth's internal behavior and related surficial manifestations have been summarized in certain standard texts on the Earth (See, e. g., Aubouin, 1965; Beloussov, 1962; Cailleux,

1969; Cook and Gaskell, 1961; De Sitter, 1956; Dunbar, 1966; Garland, 1966; Gaskell, 1967; Gutenberg, 1951; Holmes, 1965; Jacobs, 1963; Jacobs et al., 1959; Jeffreys, 1962; King, 1962; Lee, 1965; Macdonald and Kuno, 1962; Markowitz and Guinot, 1968; Matsumoto, 1967; Orlin, 1966; Poldervaart, 1955; Rikitake, 1966; Rittmann, 1962; Runcorn, 1962; Scheidegger, 1963; Shimer, 1968; Steinhart and Smith, 1966; Stevenson, 1962; Vening Meinesz, 1964).

When the Earth is considered as a global entity, the results of the above analyses indicate that most past and present geo-phenomena are manifestations of the Earth's internal behavior during the past 4.6 b. y., but that the geometry and mechanics of the causative internal behavior and the nature of the ultimate driving mechanism remain to be defined. It is the writer's hypothesis that the Tectonospheric Earth Model, derived from a dual primeval planet hypothesis (Tatsch, 1972a) provides some answers for the undefined geometry and mechanics and for the nature of the Earth's internal driving mechanism.

Chapter 4

PLATE TECTONICS, OMNIDUCTIVE PROCESSES, AND SEISMOTECTONOMAGMATIC BELTS

It has been known for some time that copper deposits are often found near plate boundaries and near other surficial vestiges of the Earth's internal behavior. One of the purposes of this chapter is to attempt to generalize this observation into a long-lived, deep-seated concept that may best be expressed in terms of "omniductive processes" and "seismotectonomagmatic belts". This concept forms one of the bases for using the Tectonospheric Earth Model as a supplementary tool in the exploration for copper.

Plate Tectonics, Copper Deposits, and the Tectonospheric Earth Model.

Before considering the applicability of the Tectonospheric Earth Model as a supplementary tool in the exploration for copper deposits, it is well to consider some of the implications of the plate-tectonics concept and to review briefly how the Tectonospheric Earth Model interprets this concept and its implications in the exploration for copper deposits.

Specific analyses regarding the global-tectonics concept include those by Audley-Charles (1972, 1975), Burrett (1974), Chesworth (1972), Fitch and Scholz (1971), Halbouty (1970), Heirtzler et al. (1968, 1973), Hey et al. (1972), Karig (1970, 1971), Klemme (1971), Kerrow and Ziegler (1972), Milsom (1970), Sclater and Francheteau (1970), Sclater et al. (1971a, 1971b), Solomon and Griffiths (1972), Tatsch (1969, 1970, 1972a), Tatsch and Talbot (1971), Uyeda and Ben-Avraham (1972), Weeks (1971), and Whiteman et al. (1975).

Many of the individual processes that are embodied in the simple plate-tectonics concept are derivable from the more general concept of the Tectonospheric Earth Model (Tatsch, 1972a). However, by going several steps farther than does the popular plate-tectonics concept, the Tectonospheric-Earth-Model concept can predict features within the Earth's substructure that are not predictable by the simple plate-tectonics concept.

The plate-tectonics concept leaves several unanswered questions, both from the standpoint of a general tectonic hypothesis and in its applicability to mineral deposits. Some of these questions have been discussed (See Tatsch, 1972a and 1973a for summaries of these discussions). Some of the quessions that plate tectonics leaves unanswered concerning the origin, evolution, and present characteristics of mineral deposits may be repeated here.

Why, for example, are some of the plate boundaries convergent while others are divergent? Why don't the plates just stay where they are? What drives some of them to the right and others to the left; some up and some down? Does a descending plate really have the sharpness and the other requisite capabilities to cut or dig its way to a depth of 830 km, or more, into fairly solid bedrock? What causes this sharpness that is not obvious either in the field or in the laboratory when we see these plate edges before they start down? Does the descending edge become sharper and sharper as it descends deeper and deeper and becomes hotter and hotter and more compressed, through some process of high-pressure, high-temperature "auto-tempering" that is not now known by metallurgists and other scientists? If such processes do not exist, at depths of 830 km and deeper, to permit these descending plates to dig their respective ways into the mantle, then is there an alternative explanation for the origin, evolution, and present characteristics of the "trenches" at the surface of the Earth (and, by induction, of their causal mechanisms, at depths of 830 km and more)? If there is an alternative explanation, what is it? Could it be that the basis for these deep trenches could have existed within the Earth ever since the Earth began about 4.6 b. y. ago? If so, then what was this primordial basis for the creation and perpetuation of these trenches, at various locations on the surface of the Earth, during the 4.6 b. y. that the Earth is believed to have been in existence?

Many of these questions remain unanswered. Those that have a bearing on the origin, evolution, and present characteristics of copper deposits are considered in subsequent sections of this chapter.

The Present Plate-Tectonics Concept as a Supplementary Tool in the Exploration for Copper Deposits.

The origin, evolution, and present characteristics of copper deposits have defied attempts to define them in terms of a single mechanism that has been operating within the Earth during the 4.6 b. y. that the Earth is believed to have been in existence. Several years ago, the plate-tectonics concept received some attention in this connection. But it is now generally agreed by explorationists that the plate-tectonics concept, per se, does not provide the final answer. There are several reasons for this deficiency of the plate-tectonics concept; and it is well to list here some recent observations that have a bearing on the geometrical, mechanical, thermal, and chemical aspects of the Earth's behavior that appear to be at variance with the predictions of the plate-tectonics concept:

a. Several of these variances may be seen in the Transverse Ranges Province of California. As a local complication of the predictions of the plate-tectonics concept, the Transverse

Ranges Province remains a severe problem for any interpretation of that area (See, e. g., Baird et al., 1974). In the Transverse Ranges Province, as in some other areas of the world, the field observations do not seem to agree with the predictions of the plate-tectonics concept.

b. Some of the data from Iceland are not reconcilable with the concept of plate tectonics (See, e. g., Piper, 1973b). Whatever may be the causal mechanism for the preferential flow of heat beneath Iceland, that region does not exhibit the self-perpetuating tendency shown by most plate margins. Although the plates may have considerable strength, considered exceptional in that region because of the great crustal thickness, the causal mechanism beneath Iceland has been sufficient to fracture the plate and to relocate the axis of preferential heat flow. Whatever this causal mechanism may be, it appears to be driving the plates apart, but not in the simple manner predictable from the basic plate-tectonics concept.

c. Very little is known about the causal mechanism for vertical motions within the Earth. Yet, these vertical motions have been of primary importance in the formation of many continental basins and geosynclines as well as in the elevation of plateaus from whence the basin-filling sediments have been derived. Whether the vertical motion is upward or downward, the causal mechanism to supply the necessary energy remains undefined (See, e. g., Osmaston, 1973; Sclater et al., 1975).

d. Pankhurst (1974) concluded that four late Caledonian granites from northeast Scotland are not related to a subduction zone. Rather, they appear to have been derived from a deep, underlying crustal layer that is not predictable from the plate-tectonics concept.

e. Earthquakes in the Himalayan region show thrust faulting. Mechanism solutions suggest that the lithospheric blocks of the Indian and Eurasian plates are underthrust toward each other. This conclusion is partially substantiated by the local geology and by the V-shaped distribution of foci in the Hindu Kush region (See, e. g., Rastogi, 1974).

f. Besides a lateral mineralogical zonation, the sub-oceanic lithosphere, as interpreted by the plate-tectonics concept, should display also a vertical variation in its chemical composition (See, e. g., Bottinga, 1974).

g. If the plate-tectonics concept, with plate creation at the ocean ridges and plate thickness of about 75 to 100 km, is correct, then one should be able to detect a sizable injection of thermal energy at the oceanic ridges into the oceans (See, e. g., Bottinga, 1974).

h. Plate-tectonics-type subduction seems not to have existed earlier than the Late Precambrian or the Early Phanerozoic within the Appalachian-Caledonian and Uralian orogenic belts (See, e. g., Cady, 1973).

i. The predictions of the plate-tectonics concept seem to be somewhat deficient along many continental margins. This is particularly true in the case of the Pacific-type margins (See, e. g., Beck and Lehner, 1974). According to the plate-tectonics concept, the Pacific-type margins are zones of collision and consumption of crustal plates. This may be true in many cases, but it does not appear to explain all the observational evidence. For example, the plate-tectonics concept does not explain fully the pronounced vertical uplifting of island arcs and the collapse of the entire area between such arcs and the continent. The widely observed repeated phases of diastrophic events are difficult to reconcile with a process of steady underthrusting and consumption of oceanic plates alone. The conspicuous ubiquity of similar structural elements related to distinct orogenic phases along Pacific as well as Alpine ranges suggests an additional driving mechanism within the orogenic system itself. In the western Andes, for example, the Cretaceous tectonics, sedimentation, and volcanism were characterized by oscillatory vertical movements of strips of basement bounded by major shear belts (See, e. g., Myers, 1974).

j. Some evidence suggests that the west coast of the U. S. did not always present the characteristics of a Pacific-type margin. During the Ordovician, for example, southwestern Nevada resembled the Atlantic-type margin (See, e. g., Hoyt and Scholle, 1974). This suggests that the characteristics of continental margins may have changed several times during the Earth's existence, depending upon whether they are near to or far from a global wedge-belt of activity (Tatsch, 1972a: page 269). According to the Tectonospheric Earth Model, the continental margins that lie above, or near, active wedge-belts of activity have the characteristics described as Pacific-type margins; those that are far removed from active wedge-belts of activity have the characteristics described as Atlantic-type margins.

k. The plate-tectonics concept has provided an interpretation for many of the observations regarding mountain building but it leaves many other observations unexplained. It is well to summarize some of those observations that have not lent themselves to interpretation through simple plate-tectonics predictions:

(1). The belts of folds and thrusts in the Earth's mountain chains are not readily interpretable in terms of the simple plate-tectonics concept.

(2). Particularly in the case of "non-collision" systems, such as the North American Cordillera, the simple plate-tectonics concept has left many unanswered questions (See, e. g., Coney, 1973b).

(3). The plate-tectonics concept leaves unsettled the debate over the regional compression versus vertical uplift and the associated gravity tectonics (Coney, 1973b). This suggests that the actual causal mechanism must embody some sort of vertical motion (Tatsch, 1972a).

(4). Coney (1973b) has suggested that the North American Cordillera and similar systems are "compressive" and that they develop as complex "strain patterns" on the leading edges of actively-driving "continental" plates as they override zones of subduction dipping beneath them. He feels that the timing and intensity of compressive strain in these belts are governed by the kinematics of the driving plate itself and that they are minimally affected by the motions of the subducted "oceanic" plate. However, Coney leaves undefined the geometrical, mechanical, thermal, and chemical aspects of the driving mechanism that produces the "kinematics of the driving plate".

1. The explanation of orogenies in terms of "colliding continents" leaves some unanswered questions. The Late Paleozoic orogeny in southeastern North America may be cited as a case in point. The salient difficulties may be summarized (See, e. g., Bowen, 1974):

(1). There is a near absence of Late Paleozoic igneous events, especially volcanism, in that area, which is presumed to have been associated with a subduction zone at that time.

(2). Stringent geographic constraints are placed upon the positioning of orogenic activity that deforms the Ouachita-Arbuckle-Ardmore region and the Marathon but leaves the Llano region unaffected.

(3). It is difficult to account for the stratigraphic equivalents of the more than a million km^3 of clastics that were placed on the North American cratonic shelf and its borderlands during the Carboniferous. The highlands supplying these clastics presumably eroded on both sides, but volumetric equivalents of the observed North American clastics are not discernible in Europe, Africa, or South America.

m. Some active areas of the Earth's surface do not lend themselves to an easy interpretation in terms of the plate-tectonics concept. For example, some Mediterranean orogenic belts appear to have formed by uplift and gravity tectonics alone, without any association with plate margins (See, e. g., Alvarez, 1973). There is no convincing evidence for a subductive origin for the Apennine allochthon. In fact, of all the mountain belts surrounding the western Mediterranean, only the Alps show convincing evidence of crustal shortening (Alvarez, 1973). Similar complexities remain unexplained in other parts of the world (See, e. g., Haile, 1973; Kent, 1973).

No attempt has been made to list all the problems that have been left unresolved by the plate tectonics concept. Rather, an attempt was made to list enough of them to reveal their scope and nature, with particular attention to those that might have a bearing on those geometrical, mechanical, thermal, and chemical aspects of the Earth's internal behavior that have influenced the origin, evolution, and present characteristics of the copper deposits.

Additional examples appear in the current literature (See, e. g., Anderson et al., 1974; Audley-Charles, 1975; Barazangi et al., 1975; Berg et al., 1975; Bowen, 1974; Bowin, 1974; Colley and Warden, 1974; Elston and Gromme, 1974; Katz, 1974; Miyashiro, 1973; Levitte et al., 1974; Meyerhoff, 1974; Purser, 1973; Sclater et al., 1975; Tschanz, 1974; Winterer, 1974; Van Breemen et al., 1974; Verma and Mital, 1974).

The Migration of the Patterns of the Surficial Manifestations of the Earth's Internal Behavior.

Very little is known about the details of the geometrical, mechanical, thermal, and chemical aspects of the Earth's internal behavior. However, it is suspected by many observers that the details of the Earth's internal behavior are best analyzed in terms of the panoramic changes that have occurred in the geophenomena and features at the Earth's surface. Ostensibly, the Earth's internal behavior is reflected in the patterns that form at the Earth's surface. Even more is revealed when these surficial patterns are observed to migrate within a spatio-temporal frame that extends through a span of billions of years.

It is well, therefore, to summarize what is known about the migration of the patterns of the surficial manifestations of the Earth's internal behavior:

a. Bowin (1974) has reported a pattern of plate motion that has migrated from the South Pacific into the Indian Ocean during the Cretaceous and Early Tertiary, followed by a migration across the Indian Ocean during the Tertiary.

b. Some observers feel that the opening of the Gulf of Aden may be a more recent and extended surficial manifestation of this migration.

c. Based on observational evidence, these migrations comprise the episodic formation of new lineaments along which sea-floor spreading and other omniductive phenomena occur (See, e. g., Sclater and Fisher, 1974).

d. In many cases, these new lineaments of seismo-tectonomagmatic activity are both translated and rotated from erstwhile spreading lineaments or axes (Tatsch, 1972a; page 739; Bowin, 1974).

e. In most of these cases, the development of the new lineaments or axes seems to follow deep-seated, long-lived patterns that might have existed deep within the Earth for billions of years.

f. These and other observations from all parts of the Earth suggest that the surficial manifestations of the Earth's internal behavior extend far beyond the current, or even the recent-past, boundaries of the Earth's surficial plates (See, e. g., Bowin, 1974).

A closer scrutiny of the above suggests that many questions remain unanswered. Why, for example, does the basic plate-tectonics concept, which appears so versatile when considered at the global and regional level, fail at the local level? Does this mean that our observations are inadequate? Or does it mean that the basic plate-tectonics concept is not the final answer?

Global Seismotectonomagmatic Belts.

A close scrutiny of the Earth's surface reveals that certain elongate belts, characterized by varying degrees of seismic, tectonic, and magmatic activity, have existed in various regions of the Earth during most of the 4.6 b. y. that the Earth is believed to have been in existence. These belts may be referred to as seismotectonomagmatic (STM) belts.

Much evidence suggests that all of these STM belts are similar, regardless of when and where formed (See, e. g., Marlow, et al., 1973) who have considered island arcs, orogenic belts, and certain other tectonomagmatic episodes. Other recent studies of various facets of the geometrical, mechanical, thermal, and chemical aspects of some of the seismotectonomagmatic belts of various ages and in various parts of the world, have been made by Hawkins (1973), Malpas, et al. (1973), Vogel (1973), Westerman (1973), Bickford and Van Schmus (1973), Justus and Roper (1973), Wielchowsky and Gilbert (1973), Miyashiro (1972, 1973), Fischer (1973), and others listed in Tatsch (1972a: chapter 7).

The Earth's Earliest Precambrian Activity as Interpreted by the Seismotectonomagmatic Belt Concept of the Tectonospheric Earth Model.

It is normally presumed by the plate-tectonics proponents that, prior to about 2.7 b. y. ago, the outer part of the Earth was so mobile that no plate-tectonics system operated (See, e. g., Burke and Dewey, 1973). This conclusion by the plate-tectonics proponents serves to emphasize some of the basic differences between the simple plate-tectonics concept and the longer-lived, deeper-seated seismotectonomagmatic-belt concept of the Tectonospheric Earth Model:

a. The longer-lived seismotectonomagmatic-belt concept of the Tectonospheric Earth Model is penesyngenetic with the origin of the Earth (See, e. g., Tatsch, 1972a: chapter 7). Thus, this concept extends farther back in time, by 70%, than does the simple shorter-lived plate-tectonics concept.

b. The seismotectonomagmatic-belt concept of the Tectonospheric Earth Model is deep seated. The "cones of activity" and the "wedge belts of activity" (i. e., the "mobile" parts of the Earth) extend to depths of at least 1000 km. The mobile parts of the Earth's tectonosphere (upper 1000 km in this model) are based on a "quasi-fixed" system that extends even deeper than 1000 km into the Earth's subtectonosphere (Tatsch, 1972a: chapter 6).

c. The plate-tectonics proponents have invoked a "permobile regime" that preceded the plate-tectonics regime and provided an approximately one b. y. period of high heat generation during which large volumes of basaltic and sialic materials differentiated from the mantle through rapid "convective circulation", probably in two stages (Jakes, 1973). The widespread preservation of very low grade assemblages in the earliest greenstone belts presents a problem to the plate-tectonics concept. In contrast, the origin, evolution, and present characteristics of these and related assemblages are predicted by the seismotectonomagmatic-belt concept of the Tectonospheric Earth Model (Tatsch, 1972a: chapter 7).

d. The known Archean volcanic belts (c 2.7 to 3.1 b. y. ago) embody petrogenic suites that are reminiscent of Cenozoić island arcs, suggesting that the oldest known (seismotectono ?) magmatic belts could have formed at sialic-oceanic interfaces as interpreted by the plate-tectonics concept (See, e. g., Goodwin, 1973: page 1066). But it is perhaps more likely that the earliest of the belts formed according to a concept that did not depend upon the collision of lithospheric plates. The seismotectonomagmatic-belt concept of the Tectonospheric Earth Model provides such a system.

e. The global distribution pattern of Precambrian sedimentary iron formations, in general, and those of the early Proterozoic (c 1.9 to 2.1 b. y. ago), in particular, suggest that these formations were deposited generally subparallel to modern oceanic and other omniductive belts of the Earth's surface (See, e. g., Tatsch, 1973a; Goodwin, 1973). The origin, evolution, and unique patterns of these iron formations can be explained in terms of the plate motions that affected the Proterozoic crust, as predicted by the plate-tectonics concept. If so, then the apparent subparallelism in the distribution patterns of the early iron formations and modern plate boundaries may be explained in terms of: (1) the Hurley-Rand (1969) concept postulating a stability for "pre-drift" continental nuclei; or (2) the evolution of the Earth's mineralization belts according to the seismotectonomagmatic-belt concept of the Tectonospheric Earth Model (Tatsch, 1972a; 1973a). In either case, it appears that the Earth's earliest metallogenic belts were formed according to some geometrical, mechanical, thermal, and chemical aspects of the Earth's internal behavior that have been operating over a long period of time, dating back perhaps 4.6 b. y., well back beyond the first appearance of lithospheric plates on the surface of the Earth.

f. Many of the aspects of the simple plate-tectonics concept are derivable from (and included within) the seismotectonomagmatic-belt concept of the Tectonospheric Earth Model. Stated in another way, the plate-tectonics concept may be considered to be a shorter-lived, shallower-seated subsystem of the seismotectonomagmatic-belt concept of the Tectonospheric Earth Model that has been operating at the surface of the Earth ever since the upper parts of the tectonosphere became stabilized about 2.7 b. y. ago (Tatsch, 1972a: chapter 18). Earlier than that, the conclusions of the simple plate-tectonics concept are based largely upon "reasonable evolutionary analyses" (See, e. g., Goodwin, 1973). Consequently, unequivocal answers about the Earth's early Precambrian are not likely to be forthcoming from the relatively short-lived plate-tectonics concept. Certain petrogenic patterns within the later Precambrian crust are compatible with some form of plate motion. But this does not preclude the application of some other model to explain those patterns (See, e. g., Goodwin, 1973: page 1047). Thus, the seismotectonomagmatic-belt concept of the Tectonospheric Earth Model encompasses all of the Earth during the 4.6 b. y. the Earth is believed to have been in existence. This contrasts with the simple plate-tectonics concept which encompasses only the surface of the Earth during the last half of the Earth's existence. This difference arises primarily from the fact that the seismotectonomagmatic-belt concept of the Tectonospheric Earth Model embodies a well-defined, long-lived, deep-seated driving mechanism (Tatsch, 1972a: page 257; 1973a: page 19). The plate-tectonics concept does not embody such a well-defined driving mechanism.

It is essential that the reader be aware of these, and other, differences (as well as some similarities) when considering the origin, evolution, and present characteristics of the Earth's copper deposits according to the seismotectonomagmatic-belt concept of the Tectonospheric Earth Model. This is particularly important when considering the present distribution of copper deposits as a function of the Earth's behavior during the past 4.6 b. y.

The Evolution of Australia as Interpreted by the Seismotectonomagmatic-Belt Concept of the Tectonospheric Earth Model.

Although devoid of collision-type orogenic belts, Precambrian Australia displays two kinds of belt: (1) miotectonic or miogeosynclinal; and (2) thermally-induced.

The latter type of belt, as represented in the Fraser and Musgrave belts, is of particular interest because it bears no resemblance to the types of belt drivable from the plate-tectonics concept (See, e. g., Rutland, 1973). Attributable to selective heat flow in narrow zones of the lithosphere, this type of belt is easily derivable from the seismotectonomagmatic-belt concept of the Tectonospheric Earth Model (Tatsch, 1972a). Closer scrutiny suggests that perhaps all of Australia could have evolved by the type of orogenic-cratonic evolution predicted by the Tectonospheric Earth Model or by the periodic cratonization in chelogenic cycles (See, e. g., Muehlberger, 1971).

Summarizing, the evolution of Australia serves to emphasize several salient differences between the predictions of the simple plate-tectonics concept and those of the seismotectonomagmatic-belt concept of the Tectonospheric Earth Model:

a. The simple plate-tectonics concept does not appear to be applicable to the Precambrian, at least not in toto, whereas the seismotectonomagmatic-belt concept of the Tectonospheric Earth Model applies to practically all of the 4.6 b. y. that the Earth is believed to have been in existence.

b. The simple plate-tectonics concept does not explain the development of the intra-continental belts such as the Musgrave and Fraser belts, whereas the seismotectonomagmatic-belt concept does.

c. The occurrence of apparently miogeosynclinal successions such as the Adelaidean are difficult to explain in terms of the simple plate-tectonics concept, but these types of succession are easily derivable from the seismotectonomagmatic-belt concept.

d. The metallogenic provinces are difficult to explain in terms of the simple plate-tectonics concept but are easily derivable from the seismotectonomagmatic-belt concept.

The Development of the Earth's Earliest Orogenic Belts as Interpreted by the Seismotectonomagmatic-Belt Concept of the Tectonospheric Earth Model.

The development of the Earth's earliest orogenic belts is not easily explained in terms of the simple plate-tectonics concept. For example, when the geometrical, mechanical, thermal, and chemical aspects of the Pan-African belts are considered, there is a strong division of opinion between two extreme possibilities (See, e. g., Hurley, 1973):

a. In-situ orogeny. In this concept, the Pan-African orogenic belts were formed in-situ by the upwelling of hot material from the mantle with accompanying subduction effects but little or no oceanic seaway or motion of the adjacent cratons.

b. Pacific-type orogeny. In this concept, the Pan-African orogenic belts developed in a manner similar to what is observed along the Pacific Ocean coasts today, i. e., sea-floor spreading, subducting belts, and the emplacement of new material by the partial fusing of the subducted material.

To some observers, these two concepts are diametrically opposed to each other (See, e. g., Hurley, 1973: page 1083).

Closer scrutiny shows, however, that a single model can produce all the surficial aspects of each of these two supposedly different types of orogeny. This model is the writer's Tectonospheric Earth Model (See, e. g., Tatsch, 1972a: chapter 3).

Some of the specific aspects of the Pan-African orogeny that are explained by the seismotectonomagmatic-belt concept of the Tectonospheric Earth Model may be listed:

a. Parts of these belts extend through parts of almost all major regions of the present continental crust (See, e. g., Hurley, 1973).

b. The geologic activity that produced this Earth-encircling orogenic system is definable within a specific period of time.

c. If one starts at any point on one of these belts, one can trace it without a break to all other belts by a continuity of foliated rocks, thermal effects, intrusive igneous material, and related structo-tectonic features.

d. There appear to be within these belts clearly defined periods of igneous and metamorphic activity.

e. The post-kinematic emplacements of granite within some of these belts are younger than the principal thermotectonic activity at a given location within a belt.

f. Some of these belts are marked by a predominance of granitoid rocks, including paragneisses, granite gneisses, and plutonic rocks in the granodiorite range and a definite scarcity of volcanic rocks relative to what might be expected in the modern-day Pacific-type orogenic belts. In these belts, ophiolite suites are more abundant than they are in provinces of earlier Precambrian age, but they are much less abundant than they are in the younger Alpine and Pacific orogenic belts.

g. Most of these belts contain relicts of ancient crust, remobilized and refoliated basement rocks, micro-continents, and large angular blocks of continental crust.

h. These belts appeared to form along great circles and to intersect orthogonally (Hurley, 1973: page 1086).

i. Thermal (and perhaps associated seismic) activity within these belts was similar to what is observed today. But some evidence suggests that the Earth's earliest thermal activity was greater (and its seismic activity was less) than it is today.

Similar analyses have been made for other continents. Some of these are simpler to analyze; e. g., North America (Tatsch, 1972a: chapter 7). Others are more complex; e. g., Eurasia (Burrett, 1974).

The Development of the Earth's Sialic Crust as Interpreted by the Seismotectonomagmatic-Belt Concept of the Tectonospheric Earth Model.

It has long been known that some mechanism capable of deforming the granitic crust has been operating for more than 3.8 b. y. (See, e. g., Black et al., 1971). Closer scrutiny suggests that at least four different styles of global tectonics have affected the Earth's surface. The last of these four styles is now described by the simple plate-tectonics concept. It is well to summarize what is known about the three styles of Earth behavior that preceded that embodied within the simple plate-tectonics concept.

Sutton has referred to the styles of the Earth's behavior as chelogenic cycles (Sutton, 1963). Under this concept, the origin, evolution, and present behavior of the Earth's surface may be described in terms of the 4 chelogenic cycles during each of which a new system of mobile belts formed and subsequently decayed to leave a progressively-larger stable area or belt as each of these chelogenic cycles followed the previous one (Sutton, 1963, 1970).

Considered chronologically, Sutton's chelogenic cycles may be described:

Stage 1: about 4.5 to 3.4 b. y. ago. Production of a granitic crust comprising many intrusions and containing no more than a few small isolated fragments of recognizable non-granitic rocks.

Stage 2: about 3.4 to 2.7 b. y. ago. Development of granitic crust that was intersected by numerous large belts of volcanic and sedimentary rocks many km thick and hundreds of km long. Vestiges of these belts remain today, but they have been folded and metamorphosed since first fromed. The oldest known granulites date from this stage.

Stage 3: about 2.7 to 2.0 b. y. ago. The Earth's surficial behavior during this stage comprised linear mobile belts that intersected stable blocks. The latter were able to support gently inclined non-metamorphic successions. Extensive fracture systems hundreds and thousands of km long and partly filled by basic intrusions formed the bases for the mobile belts at this stage.

Stage 4: about 2.0 b. y. ago to the present. Production of granitic crust divided into mobile belts and stable plates. The simple plate-tectonics concept is useful to explain the dyke swarms, rift valleys, and new ocean floors that developed during this last stage. The crust was able to support mountain chains high enough to produce molasse deposits after uplift of the sediments within the mobile belts.

An analysis of the geometrical, mechanical, thermal, and chemical aspects of these cycles suggests that any satisfactory hypothesis for the development of the Earth's sialic crust must answer certain basic questions:

a. How do we account for the original development of the Earth's sialic crust during the first 15% of geological time, i. e., within about 0.8 b. y. after the Earth formed?

b. How do we account for the increasing rigidity of the Earth's crust as it aged?

c. How do we explain the periodic occurrence of regional metamorphism, partial melting, and intense deformation within mobile belts that formed in crustal blocks that had been quiescent for hundreds of millions of years?

Answers to these questions are not known in any quantitative detail. However, the causal mechanism for the Earth's behavior during the past 4.6 b. y. must answer these and related questions. Any satisfactory causal mechanism must give particular attention to the answers to these three questions during the Earth's four chelogenic cycles. This requires a definition of the geometrical, mechanical, thermal, and chemical aspects of the Earth's deep-seated behavior during the entire span of 4.6 b. y. that the Earth is believed to have been in existence. It was with this in mind that the writer developed his "tectonospheric Earth model" as an approach to the problem of defining the Earth's internal behavior in a manner that would account for the surficial behavior during the past 4.6 b. y. One of the primary purposes of the Tectonospheric Earth Model was to define the specific geometrical, mechanical, thermal, and chemical aspects of the Earth's internal behavior that caused long-term fluctuations in the way the crust reacted to the Earth's internal behavior (Tatsch, 1959 et seq., particularly 1972a).

Considered in terms of the Tectonospheric Earth Model, it seems that, during the early stages of the Earth's existence, the primary manner in which the granitic crust was moved horizontally was not necessarily by the transport of rigid masses of continental crust but by the intense deformation of sialic crust. As time passed, successive phases of crustal dilation became more and more prominent during each chelogenic cycle. Thus, the restricted dolerite swarms of 3 b. y. ago led ultimately to the present-day arrangement of dyke swarms, rift valleys, and ocean floors.

Mountain Belts as Interpreted by the Seismotectonomagmatic-Belt Concept of the Tectonospheric Earth Model.

Mountain belts, when considered under the seismotectonomagmatic-belt concept of the Tectonospheric Earth Model, are caused by a single driving mechanism (Tatsch, 1972a: chapter 12). This concept appears to differ, in that respect, from the conclusion of the simple plate-tectonics concept that there are two types of mountain belt:

a. Subduction-type mountain belts are developed over a subduction zone where the oceanic plate is consumed. If the non-consuming plate is oceanic, then the mountain belt is an island arc; if it is continental, then the mountain belt is Cordilleran.

b. Collision type mountain belts are developed from the collision of (1) two continental plates, or (2) a continental plate and an island arc.

Closer scrutiny shows that there are many orogenic belts that fall into neither of the two types defined by the simple plate-tectonics concept. Observational data seem to imply that some of these belts were formed by a process that involved:

a. Little or no relative motion of crustal plates.

b. No appreciable crustal shortening between the plates.

c. No former oceanic plates that have been consumed.

One example of this type of orogenic belts may be seen in the Precambrian of Africa (See, e. g., Shackleton, 1973). Paleomagnetic evidence from Africa suggests extensive and complex motions relative to the magnetic dipole axis (McElhinney et al., 1968). But motion between the African cratons does not seem to have occurred during the past 930 m. y. (Briden et al., 1971).

Based on these and similar cases from other continents, it appears that many ancient orogenic belts may be the result of processes that do not involve either plate motion or melting caused by subductive "frictional" heat along a Benioff plane (Shackleton, 1973). If this is the case, then there are actually three types of orogenic belt, i. e., the two predicted by the simple plate-tectonics concept and those produced by "other processes".

A detailed discussion of this matter is beyond the scope of this analysis. Suffice it to say here merely that all mountain belts, regardless of differences at the surface of the Earth, share a common causal mechanism when analyzed according to the seismotectonomagmatic-belt concept of the Tectonospheric Earth Model (Tatsch, 1972a: chapter 12).

The Development of Faunal Assemblages as Interpreted by the Seismotectonomagmatic-Belt Concept of the Tectonospheric Earth Model.

It has been known for some time that certain faunal assemblages have developed penecontemporaneously in widely separated areas. For example, the Aphelapsis appeared relatively suddenly in shelf and miogeosynclinal sites of both eastern and western North America (Lochman-Balk, 1974). Simultaneously, there was an abrupt extinction of the established Crepicephalus faunal assemblage in both areas.

This suggests that this faunal change was caused by a relatively rapid change in one or more physical factors of the marine environment. This is predicted by the seismotectonomagmatic-belt concept. That is, the shelf and miogeosynclinal conditions were very similar in the surficial manifestations of the Aleutians-to-Galapagos seismotectonomagmatic belt (western U. S.) and in those of the Galapagos-to-Gibraltar seismotectonomagmatic belt (eastern U. S.). Thus, the abrupt environmental change in these two belts could have been, for example, a near-simultaneous, rapid cooling of the ocean waters over a relatively short span of geologic time, as predicted by the STM-belt concept (Tatsch, 1972a).

Continent-Continent Collisions as Interpreted by the Seismotectonomagmatic-Belt Concept of the Tectonospheric Earth Model.

As earlier indicated, the STM-belt concept eliminates many of the problems encountered by a close scrutiny of the basic plate-tectonics concept (See, e. g., Tatsch, 1972a: page 821; 1973a: page 65). It is beyond the scope of this book to consider the details of all such regions of the Earth. It is well, however, to consider briefly one such system, i. e., the extension of the Appalachians into the Ouachita-Marathon system (See, e. g., Griffin, 1974; Keller and Cebull, 1974). This system does not appear to confirm the possibility of a late Paleozoic continent-continent collision on the Appalachians as proposed by Dewey and Bird (1970).

The seismotectonomagmatic-belt concept resolves this enigma by considering the spatio-temporal relations between the geometrical, mechanical, thermal, and chemical aspects of the past development and present behavior of the interior and exterior zones of orogens (See, e. g., Tatsch, 1972a: chapters 7, 10, and 12). Two questions raised by Griffen (1974) may be cited:

a. In continent-continent collisions, how may stresses be transmitted from the margins of impinging continental plates into areas near the foreland and produce deformation in the miogeosyncline, but at the same time leave relatively untouched the intervening regions (e. e., the central portions of the inner Piedmont belt in the southern Appalachians)?

b. In the absence of continent-continent collision, how may we rationalize the miogeosynclinal deformation, which lags 90 to 100 m. y. behind thermal-core ages?

Griffin (1974) and Keller and Cebull (1974) leave these questions unanswered. Obviously, however, these and related questions must be resolved in any global concept that hopes to provide an explanation for the past development and present behavior of the orogenic belts on the surface of the Earth. The writer nominates the STM-belt concept for this purpose, and he suggests that the interested reader consult the applicable portions of his discussion of the related problems (See, particularly, Tatsch, 1972a: chapters 7, 10, and 12; 1973a: chapters 2 and 3. Figure 7-7-0-2, on page 341 of the initial reference appears significantly applicable to this problem area).

Descending Lithospheric Plates as Interpreted by the Seismotectonomagmatic-Belt Concept of the Tectonospheric Earth Model.

De Bremaecker (1974) has recently analyzed the phenomena that occur during the hypothesized descent of lithospheric plates through the upper mantle (Minear and Toksoz, 1970; McKenzie, 1972; Griggs, 1972). De Bremaecker emphasizes the necessity for considering these phenomena as parts of complete systems that are both kinematically and dynamically satisfactory. Because the temperature and convective motions are very strongly coupled, it would be equally as false to say that the temperature determines the motion as to say that the motion determines the temperature (De Bremaecker, 1974: page 1). When the temperature in a convecting upper mantle is taken into consideration, some of De Bremaecker's models suggest markedly lower temperatures than those generally proposed when using less-complete analyses (De Bremaecker 1974: page 10).

Under the STM-belt concept, all geometrical, mechanical, thermal, and chemical aspects of the observed phenomena are considered to a depth of roughly 1000 km and over the entire 4.6 b. y. that the Earth is believed to have been in existence. A detailed analysis of these aspects is beyond the scope of this book. However, the interested reader if referred to the results of analyses in these areas (See, e. g., Tatsch, 1972a: chapters 7, 10, and 12; 1973a: chapters 2 and 3).

Transform Faults as Interpreted by the Seismotectonomagmatic-Belt Concept of the Tectonospheric Earth Model.

Ever since Wilson (1965) proposed the transform-fault concept, there has been some confusion about the exact difference between transform faults and transcurrent faults. Hill (1973) has aided this confusion by attempting to show that the San Andreas fault is not a transform fault. As a corollary to this argument, Hill questions the plate-tectonics approach to the past development and present behavior of the structure of western North America as interpreted by McKenzie and Morgan (1969), Atwater (1973), and others.

In spite of these controversies, however, the basic transform-fault concept has proved helpful in analyzing many of the geometrical aspects of global tectonics. Unanswered are questions regarding some of the mechanical, thermal, and chemical aspects of the Earth's behavior associated with transform faults. This includes several important questions regarding a mechanism for creating the stress field that initiates transform faults. Some work has been done in this area, and some proposals have been made. It has been shown, for example, that transform faults and transcurrent faults are two distinctly different geometrico-mechanical types of strike-slip faults (See, e. g., Freund, 1974).

In one of the proposed causal mechanisms, it is hypothesized that the work involved in a ridge-transform system is smaller than is that in an oblique ridge (See, e. g., Lachenbruch and Thompson, 1972). Then, by the minimum-work principle, transform faults develop preferentially. Close scrutiny of the geometrical, mechanical, thermal, and chemical aspects of the Earth's known transform faults suggests that several basic questions remain unanswered:

a. Why, for example, do transform faults develop at some places but not at others?

b. What mechanism within a sphere can produce (and maintain) systems of orthogonal fractures on the surface of the sphere?

c. More specifically, what is the actual stress pattern that initiates (and maintains) the myriad of transform faults observed on the Earth's surface?

Faced with these and related questions about transform faults, the writer has used the seismotectonomagmatic-belt concept of the Tectonospheric Earth Model to obtain partial answers to these questions:

a. Transform faults develop wherever there is a transmittal, to the surface, of the effects of the Earth's rotary equilibration (Tatsch, 1972a: page 119 et seq., with particular attention to Figure 3-3-1, on page 120).

b. The actual stress pattern that initiates (and maintains) the transform faults at the Earth's surface is that which is transmitted by the Earth's rotary equilibration (Tatsch, 1972a: page 119 et seq.; 1973a: page 20, with particular attention to Figure 2-2, on page 21).

c. An extension of this concept may be used to explain the specific shapes of oceanic ridges that embody transform faults (Tatsch, 1972a: page 742, with particular attention to Figure 16-10-5-4, on page 744).

Inter-Arc Basins as Interpreted by the Seismotectonomagmatic-Belt Concept of the Tectonospheric Earth Model.

The geometrical, mechanical, thermal, and chemical aspects

of the tectonic evolution of inter-arc basins are not fully understood. However, it is well to summarize what is known and what has been hypothesized about these unique surficial manifestations of the Earth's internal behavior:

a. Free-air gravity-anomaly profiles "landward" of the Bonin Arc reveal a narrow belt of large-amplitude negative anomalies (about 75 km wide) superimposed on a broad belt of positive anomalies (about 500 km wide) (See, e. g., Watts, 1974).

b. The negative anomalies correlate with the Bonin Trough and are at least partly caused by substantial thicknesses of sediments underlying the trough.

c. The positive anomalies correlate with a regional rise in topography of several hundred meters.

d. Profiles "landward" of the Mariana and Tonga arcs show a broad belt of positive anomalies (about 500 km wide).

e. These positive anomalies are closely associated with the Mariana Trough and Lau Basin.

f. They may be at least partly caused by relatively dense material underlying the basins.

g. These profiles, combined with other observations, suggest a two-stage model for the tectonic evolution of inter-arc basins: (1) an initial stage of regional uplift of the plate, combined with the development of a narrow fault-bounded trough "behind" the arc; and (2) a later stage of intrusion of dense material in the region of the trough, combined with crustal extension and plate separation (See, e. g., Watts, 1974).

These observations place important constraints on the interpretations of the regional gravity field of island arcs and upon models for the evolution of these fields. One of the purposes for the development of the seismotectonomagmatic-belt concept was to provide a geometrical, mechanical, thermal, and chemical framework within which these constraints could operate to provide the evolution of the observed fields (See, e. g., Tatsch, 1972a: chapter 13; 1973a: chapters 2 and 3).

Omniductive Processes and Ultramafic Rocks in the Paired Belts as Interpreted by the Seismotectonomagmatic-Belt Concept of the Tectonospheric Earth Model.

Stevens et al. (1974) have analyzed the ultramafic rocks in the paired belts (Hess, 1955) of the Appalachians in Canada and the U. S., including most ages, as well as the belts of the western part of the North American continent, including Alaska. Stevens and his associates concluded that the ultramafic rocks in these systems are basically of 4 main types: (1) obducted ophiolites derived from oceanic lithosphere; (2) oceanic lithosphere trapped beneath island-arc material; (3) differentiated mafic sills; and (4) pyroxenite-rich plutons derived from upwelling mantle diapirs above a subduction zone.

These obductive and subductive processes recognized by Stevens and his associates are included within the omniductive processes predictable within the STM belts associated with those areas (Tatsch, 1972a, 1973a).

Gravity Tectonics as Interpreted by the Seismotectonomagmatic-Belt Concept of the Tectonospheric Earth Model.

In its broadest sense, the concept of gravity tectonics assumes that all mass displacements in the Earth are ultimately motivated by gravitational forces. Thus, all geological processes and phenomena, according to the gravity-tectonics concept, are interpretable in terms of the potential gravitational energy of the global masses involved and the expenditure of that energy. In this broad sense, the gravity-tectonics concept is of little practical use in local and regional structural problems.

In a narrower, more-practical sense, the gravity-tectonics concept implies only that a given, regionally integrated tectonic system at the crustal level has lost potential gravitative energy in the course of deformation, even though parts of the system may have gained energy (See, e. g., DeJong and Scholten, 1973).

Under the seismotectonomagmatic-belt concept, the causal mechanism for the Earth's behavior may be expressed as a function of the disequilibration energy inherent in its initial state (See, e. g., Tatsch, 1973a: page 19). In simplest terms, this disequilibration energy consists essentially of the resultant of two factors, each of which has been varying in a predictable manner during the past 4.6 b. y.: (1) the potential energy inherent in the geogenetically disequilibrated shape of the basic 5400-km primordial Earth (Tatsch, 1972a); and (2) the selectively-channeled energy from the preferential flow of heat, and possibly of volatiles, outward from the 3 mutually-orthogonal belts which are "active" along the tensile portions of the surficial traces of the 3 planes separating the solid protions of the primordial Earth into 8 octantal subtectonospheric blocks.

Within the geometrico-mechanical constraints of the model (Tatsch, 1972a: page 20 et seq.), the gravity-tectonics concept is derivable from the STM-belt concept.

During the past decade, the emphasis on major horizontal movements involved in the plate-tectonics concept has, perhaps unjustly, caused a severe lessening of the emphasis on discussions of gravitational mechanics in the crust. This shift of emphasis has placed, perhaps, too much attention on questions of "what?", "where?", and "when?", with too little attention on the questions of "why?" and "how?".

One of the purposes for the development of the Tectonospheric Earth Model (Tatsch, 1959 et seq.) was to permit the simultaneous consideration of all the geometrical, mechanical, thermal, and chemical aspects of the past development and present behavior of the Earth that might permit a systemized approach to answering simultaneously all five questions: what, where, when, why, and how?

The Observed Seismicity of the Baffin Bay Region as Interpreted by the Seismotectonomagmatic-Belt Concept of the Tectonospheric Earth Model.

In the basic plate-tectonics concept, the Baffin Bay region is considered to be within the stable North American plate. However, it is well known that this region is one of the most seismically active areas in the Canadian Arctic. This seismicity is aligned along two parallel lines about 200 km apart and striking roughly N40^{o}W (See, e. g., Qamar, 1974).

According to the STM-belt concept, the Baffin Bay region lies within the surficial manifestations of the basic Aleutians-to-Gibraltar belt. Furthermore, the N40^{o}W strike found by Qamar (1974) for the two linear trends of epicenters is roughly parallel to the Aleutians-to-Gibraltar line. Consequently, the STM-belt concept provides a means of understanding the relationship of the Baffin Bay seismicity to the tectonic forces acting in that region.

Alpine Peridotite Types as Interpreted by the Seismotectonomagmatic-Belt Concept of the Tectonospheric Earth Model.

The Alpine peridotite rocks have long been a source of

controversy (See, e. g., Steinmann, 1905; Benson, 1926; and Hess, 1955). This controversy derives primarily from (1) the recognition that these rocks are mantle derived; and (2) the confusion about their mode of occurrence.

Steinmann (1905), Benson (1926), and Hess (1955) all recognized that the Alpine peridotites occurred in linear belts of intrusive origin. But much confusion still persists on other details (See, e. g., Moores, 1973).

Under the plate-tectonics concept, ophiolites are normally presumed to have developed by a "flip" of subduction (McKenzie, 1969). The evidence cited for this "flipping" is the observed presence of ophiolite slabs, such as, for example, those of New Guinea (See, e. g., Hamilton, 1973).

Under the seismotectonomagmatic-belt concept, it is not necessary to invoke the subductive "flipping" to produce these ophiolites (Tatsch, 1972a; 1973a).

The Geologic Development of the Indonesian Area as Interpreted by the Seismotectonomagmatic-Belt Concept of the Tectonospheric Earth Model.

The exact details of the geometrical, mechanical, thermal, and chemical aspects of the geologic development of the Indonesian area are not known. But it is well to summarize what is known and what has been hypothesized about the past development and present behavior of this area:

a. Very little is known about the Precambrian and early Phanerozoic of this area of the Earth.

b. In the early Mesozoic, it is normally hypothesized that Indochina, Indonesia, and the older parts of the Philippines were located on the north edge of the hypothesized Gondwanaland

(See, e. g., Cronk, 1974).

c. During the Mesozoic, the region from the Celebes Sea westward moved to its present position.

d. During the Tertiary, the eastern islands and New Guinea moved to their present locations.

e. During the mid-Tertiary, change in direction of the motion of the Pacific plate from a northward to a northwestward direction, combined with the northward movement of Australia, caused the deformation of the formerly straight outer Banda Arc.

f. Other important developments within the area were caused by the shifting of plate boundaries, which in turn caused the transfer of plate fragments from some plates to others.

The causal mechanism for these hypothesized phases in the development of the Indonesian area is not known. This arises primarily from the fact that the basic plate-tectonics concept leaves the details of a causal mechanism largely undefined, except to hypothesize that mantle convection may be involved (See, e. g., Cady, 1974; Kidd et al., 1974; Solomon et al., 1974; Turcotte and Oxburgh, 1974). The seismotectonomagmatic-belt concept goes a step farther and defines the geometrical, mechanical, thermal, and chemical aspects of a causal mechanism that has been operating within the Earth during the 4.6 b. y. that the Earth is believed to have been in existence (See, e. g., Tatsch, 1972a: page 257; 1973a: pages 19 and 26).

When this driving mechanism of the Tectonospheric Earth Model is applied to the past development and present behavior of the Indonesian areas, the results may be summarized:

a. The Precambrian and early Phanerozoic developments are interpretable in terms of the orogenic-cratonic evolution of the Earth's land and water areas (See, e. g., Tatsch, 1972a: chapter 7; 1973a: page 43).

b. During the Mesozoic and Tertiary, the present continental land areas of the Indonesian area were pared from larger landmasses by the surficial manifestations of the geometrical, mechanical, thermal, and chemical activity within (1) the equatorial seismotectonomagmatic belt passing through that longitude and (2) one or more meridional STM belts that intersected the equatorial STM belt in the applicable areas.

c. Although the surficial manifestations of the Earth's internal behavior in this area are sigmoidal or sinusoidal, the the STM-belt concept requires that, at depth, the causal mechanism is rectilinear along mutually-orthogonal great circles. Observational evidence confirms this predicted at-depth rectilinearity of the driving mechanism (See, e. g., Johnson et al., 1974).

Mineral Deposits as Interpreted by the Seismotectonomagmatic-Belt Concept of the Tectonospheric Earth Model.

It has been known for some time that many mineralizations appear to be aligned in belts that are closely associated with plate-tectonic boundaries and with certain other tectonomagmatic, seismotectonic, and omniductive features of the Earth's surface. The exact nature and cause for this association between mineralization belts and the geologic structure and behavior of the Earth are not known. One of the purposes of this book is to make a preliminary analysis of this association between copper mineralization belts and the internal behavior of the Earth during the past 4.6 b. y. One of the premises of the book is that the features and phenomena of the Earth's surface reflect the geometrical, mechanical, thermal, and chemical aspects of the Earth's internal behavior during the 4.6 b. y. that the Earth is believed to have been in existence. By studying the behavior of the Earth's surface, one can determine, hopefully, some of the aspects of the Earth's internal behavior.

One of the features and phenomena of the Earth's surface that may reflect the geometrical, mechanical, thermal, and chemical aspects of the Earth's internal behavior is the spatio-temporal panorama presented by the origin, evolution, and accumulation of mineral deposits during the past 4.6 b. y. One aspect of this panorama reveals itself in the relationship of the local, regional, and global patterns of metallogenesis to the corresponding patterns of the development of the seismotectonomagmatic belts.

Close scrutiny of the Earth's surface suggests that local, regional, and global metallogenic zoning is produced by (1) different structural environments, (2) crust and mantle heterogeneities, and (3) different histories of evolution (See, e. g., Radkevich, 1972). Most local zoning reflects the influence of episodic intruso-extrusive activity plus the geometrical, mechanical, thermal, and chemical aspects of the surficial manifestations of the Earth's internal behavior.

Many regional metallogenic zones are constituted similarly in all parts of the world; that is, (1) gold and copper in eugeosynclines; (2) tin and tungsten in terrigenous miogeosynclines; (3) lead and zinc in geanticlines composed of limestones.

Global zoning may be seen in the circum-Pacific belt, which is separable into an inner and an outer megazone. The inner megazone, with Neogene andesite magmatic rocks and mineralization of the femic type (Cr, Ni, Cu, Au), originated close to the ocean on crust of oceanic and transitional types. The outer megazone was developed on crust of continental type, being characterized by mineralization of sialic affiliation (Sn, W) related to Mesozoic granites and mineralization of femic affiliation in eugeosynclines (North and South American Cordillera).

Other geometrical, mechanical, thermal, and chemical aspects of some of the Earth's local, regional, and global mineralogenic associations with seismotectonomagmatic belts have been analyzed by Taylor and Steveson (1972), Rundqvist (1968), Shchleglov (1968), Whiteman et al. (1975), Wiltse (1973), Maucher (1972), Laznicka and Wilson (1972), Watson (1973), Goossens (1972, 1973), and Stanton (1972).

Chapter 5

INTRUSIVE AND EXTRUSIVE ACTIVITY: A LONG-LIVED, DEEP-SEATED BASIS FOR THE ORIGIN AND EVOLUTION OF MINERAL DEPOSITS

Many intrusive and extrusive episodes appear to be comagmatic even when not contemporaneous. In some of these episodes, ultramafic rocks have been wholly or partly intruded after the main metamorphic and deformational phases of the tectonomagmatic episode. In other episodes, this sequence has been reversed. In still others, intrusive and extrusive phases have been pene-contemporaneous with the metamorphic and deformational phases. The Kenoran orogeny (2.73 to 2.33 b. y. ago) displays several distinct igneous, metamorphic, deformational, and faulting phases within the Canadian Shield (See, e. g., Weber and Stephenson, 1973). In other times and places, similar phases occur but not necessarily in this same sequence.

Intrusive and extrusive activity is closely associated spatially and temporally with tectonic activity. The intrusion of voluminous basaltic dikes in the volcanic and batholithic structures often continues before, during, and after deformation, suggesting thereby a continuous process of basic and acidic magmatism contemporaneous with tectonic activity (See, e. g., Strong, 1973). All alkali olivine basalt series are similar to other alkali olivines regardless of whether formed in deep-oceanic, island-arc, or continental environments (See, e. g., Schwarzer and Rogers, 1973).

These and other recent observations suggest that the various types of intruso-extrusive activity, as well as the associated tectonic activity, are motivated by a long-lived, deep-seated, global mechanism that has been operating within the Earth during most of the 4.6 b. y. that the Earth is believed to have been in existence. It is well to examine these and other aspects of the intrusive and extrusive episodes, in an attempt to define these episodes in terms of their constituting a long-lived, deep-seated basis for the origin and evolution of mineral deposits.

Intrusive and extrusive activity has played an important role in the evolution of the Earth's tectonosphere during the past 4.6 b. y. This intrusive and extrusive activity includes volcanic activity, the creation of basaltic traps, both in continental and in oceanic areas, mantle upwelling at crustal rifts, and the emplacement of batholiths, laccoliths, and other intrusive bodies during the past 4.6 b. y.

This book considers intrusive and extrusive eruptives as a single entity. This is done in order to emphasize the concept that a single global deep-seated, long-lived mechanism is responsible for the behavior of both of these types of eruptives. The term "eruptive" is used in its general sense (Challinor, 1964), i. e., including both intrusive and extrusive activities involved in breaking through from one level to another within the tectonosphere. Thus, an eruptive that "breaks through" to a subsurface level is an intrusive; one that breaks through to the surface, an extrusive. An intrusive during one orogeny may become an extrusive through the action of a subsequent orogeny.

Arthur Holmes pointed out many years ago that the three major problems of igneous petrology are: (1) the nature and source of primary magmas; (2) the mechanisms of magmatic intrusion and emplacement; and (3) the processes that have brought about the manifold diversity of igneous rocks. One additional problem may be added to these: (4) the nature and source of the energy for the ascensive mechanism. These four problems have been discussed elsewhere (Tatsch, 1973a: chapter 4).

Intrusive and Extrusive Activity as Interpreted by the Tectonospheric Earth Model.

Most observers agree that present mineral deposits have been derived from materials that have existed within the Earth for at least 3.6 b. y. But there is no general agreement on the evolutionary processes that have led to the localization of minerals. Many layered deposits of metallic minerals were contemporaneous with the formation of the enclosing rock. Other minerals were introduced later, perhaps through some geometrical, mechanical, thermal, or chemical aspect of those particular rocks that favored a preferential path for the mineralizing materials. Still others were redistributed from earlier mineralizations during subsequent tectonothermomagmatic activity. These redistributions, in some cases, caused enrichment; in others, impoverishment (Bain, 1968).

According to the Tectonospheric Earth Model, the origin, evolution, and present characteristics of mineral deposits are closely associated with the Earth's intrusive and extrusive activity, which, in turn, is associated with orogenic activity. These associations may be seen in the evolution of most continents.

The geological history of Africa, for example, bears a close relationship to the origin, evolution, and present characteristics of mineral deposits within that continent. The history of most continents can be studied best as an orogenic-cratonic evolution. Thus, we find that Africa has undergone at least seven major orogenic events, each of which had a profound effect on the origin, evolution, and present characteristics of the mineral deposits within the African continent: (1) Transvaalian, 3 b. y. ago; (2) Shamvaian, 2.8 to 2.5 b. y. ago; (3) Eburnian-Huabian, 1.85 b. y. ago; (4) Kilbaran, 1.1 b. y. ago; (5) Damaran-Katangan (Pan-African), 0.55 b. y. ago; (6) Acadian-Hercynian, 0.40 to 0.20 b. y. ago; and (7) Alpine, 0.18 to 0.02 b. y. ago.

In Africa, as in other continents, the earliest orogens appear to have been less affected by more recent orogenies. Therefore, the mineral deposits within the oldest orogenic belts can be expected to differ from those within more recent orogenic belts. Thus, the older orogenic belts (long since welded into the present cratons) contain Africa's principal deposits of asbestos, chromium, diamonds, gold, and iron. The younger orogens contain Africa's principal deposits of beryllium, cobalt, copper, lead, niobium, tantalum, tin, and tungsten.

According to the Tectonospheric Earth Model, all minerals that have been deposited during the past 4.6 b. y. have originated within an active seismotectonomagmatic belt (Tatsch, 1972a). Thus, Africa's primary deposits of asbestos, chromium, diamonds, gold, and iron originated within seismotectonomagmatic belts that controlled the three earliest major African orogenies, i. e., the Transvaalian, Shamvaian, and Eburnian-Huabian. Similarly, Africa's primary deposits of beryllium, cobalt, copper, lead, niobium, tantalum, tin, and tungsten became concentrated within the seismotectonomagmatic belts that controlled the four more-recent major African orogenies, i. e., the Kilbaran, Pan-African (Damaran-Katangan), Acadian-Hercynian, and Alpine.

Almost all of the Earth's oldest basic seismotectonomagmatic belts have long since become welded to the cratons (and in most cases are now considered to be part of the respective cratons). However, the Earth's present metallogenic provinces attest to the erstwhile activity of these fossil seismotectonomagmatic belts. In some cases, the process by which these erstwhile active seismotectonomagmatic belts became welded to the respective cratons has caused the original deposits within the older fossil seismotectonomagmatic belts to become less clear-cut than is characteristic within the more-recently active seismotectonomagmatic belts. Thus, minerals characteristic of the older seismotectonomagmatic belts occur where large segments of these older seismotectonomagmatic belts have been involved in younger orogenic events. Consequently, a number of asbestos, gold, and iron deposits found near the margins of younger orogens probably owe their present distribution to this later orogenic activity.

In some cases, minerals originally emplaced in fossil seismotectonomagmatic belts have been transported to deposits within younger seismotectonomagmatic belts. One example concerns the diamondiferous gravels that have been transported by the Orange River into Tertiary and Recent sediments. Similar examples may be found in other continents involving different minerals at different times.

Intrusive and extrusive activity, according to the Tectonospheric Earth Model, is associated with the "cones of activity" as well as with the seismotectonomagmatic belts of the model (Tatsch, 1972a: chapter 6). That is, intrusive and extrusive activity due to a given energy source (existing at the surface of the subtectonosphere) would be expected to occur somewhere within a cone of activity centered on the "energy source" as a vertex and intersecting the surface of the Earth in a circle with a radius of approximately 785 km (= 1000 pi/4) in the idealized case (Tatsch, 1972a: figure 6-2e-2, chapter 6).

When local cones of activity are formed for every point on the subtectonospheric fracture system, then the surficial trace of their vertexes consists of 3 mutually-orthogonal great circles, aggregating about 102,000 km in length. The resulting cones of activity form an infinity of cones whose "envelope" may be described as 3 global tectonospheric seismotectonomagmatic belts. At the surface of the Earth, where the radius is 1.18 times that at the base of the tectonosphere, the centerlines of these cones form 3 mutually-orthogonal great circles aggregating 102,000 x 1.18 = 120,000 km (Tatsch, 1972a: chapter 6).

All intrusive and extrusive activity, according to the model, is motivated by the Earth's driving mechanism. Intrusive and extrusive activity resulting directly from the Earth's driving mechanism would normally be expected to occur within the present position of an active seismotectonomagmatic belt; that resulting indirectly from the Earth's driving mechanism may occur anywhere within the Earth without regard to present positions of the active seismotectonomagmatic belts (Tatsch, 1972a).

Because, at any given time, active or effective energy sources may be expected along about 50% of the subtectonospheric fracture system (i. e., along about 51,000 km), intrusive and extrusive activity would be expected to occur within about half of the global seismotectonomagmatic belt system (i. e., along "belts" aggregating about 60,000 km on the surface of the Earth) at any one time. Intrusive and extrusive activity would not necessarily occur simultaneously in all active segments of the global seismotectonomagmatic belts. Also, some types of magmatic activity normally would not accompany certain types of tectonic activity. For example, ultrabasic effusives would not normally accompany compressive tectonics according to the model. Nor would more than one specific type of magmatism be expected to occur at a given time at a given location (Tatsch, 1972a).

Crustal and Upper-Mantle Evolutionary Modes Postulated by the Tectonospheric Earth Model.

Depending upon what basic assumptions are made, the Earth's crust and upper mantle developed, according to the Tectonospheric Earth Model, through any one of three basic modes of accretionary evolution: (1) octantal-fragment, (2) multiple-fragment, and (3) composite-fragment (Tatsch, 1972a: chapter 3). The type and behavior of magmatic activity during the past 4.6 b. y. would have been different in each of these three accretionary modes of tectonospheric evolution. These may be summarized.

In the octantal-fragment mode for the evolution of the Earth's tectonosphere, the 5 "terrestrial" octants of Earth Prime are not completely fragmentized prior to accretion onto the basic 5400-km primordial Earth (Tatsch, 1972a: chapter 3). In such a case, parts of each of the 5 terrestrial octants of Earth Prime might still exist today as unmodified portions of the Earth's present tectonosphere, perhaps even as subsurficial portions of presently-existing continental shields.

If the Earth's tectonosphere evolved through the octantal-fragment mode, the intrusive and extrusive activity during the past 4.6 b. y. would have been somewhat different from that in the multiple-fragment and composite-fragment modes for the evolution of the Earth's tectonosphere. The details of these differences have been discussed (Tatsch, 1972a), but they need not be considered further here.

The multiple-fragment mode differs from the octantal-fragment mode primarily in the degree to which the 5 "terrestrial" octants of Earth Prime were fragmentized prior to accretion onto the basic 5400-km primordial Earth. If the degree of pre-accretion fragmentation was fairly complete, then it is not very likely that any of the actual surficial rocks of present continental shields would now be identifiable with unmodified portions of the original octants of Earth Prime, because intrusive and extrusive activity during the past 4.6 b. y. would have modified these multiply-fragmentized primordial rocks.

The composite-fragment mode may be considered as an intermediate mode lying between the octantal-fragment mode and the multiple-fragment mode. Depending upon the specific degree of fragmentation assumed to have occurred within the octants of Earth Prime prior to accretion onto the basic 5400-km primordial Earth, this mode occupies one of a myriad of intermediate positions within the entire spectrum, or envelope, bounded by the octantal-fragment and multiple-fragment modes as extremes. A similar statement may be made about the modification of these fragments by intrusive and extrusive activity during the past 4.6 b. y. That is, the amount and degree of modification by intrusive and extrusive activity in the composite-fragment mode would be intermediate between that in the octantal-fragment and that in the multiple-fragment mode of tectonospheric evolution.

Intrusive and Extrusive Activity as a Function of Paths of Preferential Heat Flow within the Tectonosphere.

Within the seismotectonomagmatic belts, the nature and location of intrusive and extrusive activity depends largely upon: (1) the nature of the local stress field and (2) the paths of preferential heat flow (Tatsch, 1972a: chapter 6).

If an energy source at some point along the subtectonospheric fracture system consisted of heat, and if the tectonosphere above that point were homogeneous, then it might be a relatively simple matter to determine the most probable paths that the heat might follow through the tectonosphere to the surface of the Earth. Because the tectonosphere (and specifically the seismotectonomagmatic belt) is not homogeneous, the preferential heat-flow path (from the surface of the subtectonosphere to the surface of the Earth) might be quite circuitous in order to circumvent heterogeneities of low thermal conductivity. The exact path followed by the heat would influence the nature and location of the resulting magmatic activity, whether intrusive or extrusive. This, according to the model, accounts for the circuitous routes taken by some feeders, or conduits, to volcanoes and to intrusive bodies (Tatsch, 1972a: chapter 9).

Paths of preferential heat flow within the Earth's tectonosphere are of interest to the origin, evolution, and present characteristics of copper deposits because most paths of preferential heat flow are closely associated with paths of preferential flow of volatiles and magmas. The importance of these associations may be emphasized by considering them in connection with the origin of copper deposits.

Thus, we see that magmas appear to be the ultimate source of essentially all the constituents of copper deposits. Some of the volatile constituents (e. g., water, oxygen, and carbon dioxide) appear to have followed circuitous paths from the magmas, to the hydrosphere, and then to copper deposits, but their ultimate source appears to be the magmas.

Magma crystallization may be analyzed in three stages: (1) initial, (2) intermediate, and (3) final. Mineral deposits fall, likewise, into one of these categories, depending upon what stage of crystallization is currently applicable. Thus, many metallic oxides, sulfides, and native metals are representative of the initial stage of magma crystallization. Other minerals, generally silicic, are representative of the intermediate stage of magma crystallization. Still other minerals, highly silicic, are representative of the final stage of magma crystallization.

During the initial stage of magma crystallization, the abstraction of the early crystallization products leaves a normally silicic fluid enriched in volatiles. These tend to collect in the upper part of the magma chamber until a path of preferential escape develops, permitting them, under suitable conditions, to form contact-matasomatic mineral deposits.

Toward the close of the magma solidification, some of the highly silicic accumulated fluids may pass through paths of preferential escape to form pegmatites. These fluids are accompanied or followed by aqueous solutions containing the compounds from which the ores are deposited by replacing the pegmatitic minerals. Upon final consolidation of the magma, the residual fluids are ejected along paths of preferential flow (i. e., of negative pressure gradient). From these, nearly all epigenetic mineral deposits are formed.

As the fluid cools, the gaseous constituents form magmatic liquids. These, upon nearing the surface, form hydrothermal solutions, either with or without the addition of meteoric waters. Some alteration occurs as these solutions traverse paths of preferential flow through the various wall rocks. Many become the alkaline solutions from which ores are deposited.

The paths of preferential flow may be cracks, joints, bedding planes, rock pores, and other openings that might have been produced by the geometrical, mechanical, thermal, and chemical behavior of the Earth's tectonosphere during the past 4.6 b. y. The great variety of wall rocks involved in the various conditions of temperature and pressure determines the exact locations for the hypothermal, mesothermal, and epithermal deposits.

Subsequent processes may alter the initial deposits. For example, weathering may release minerals for transport to sedimentary basins. Organic processes contribute to other alterations, including the conversion of plants and animals to fossil fuels. Some subaerial oxidation processes convert ore minerals into more usable forms, others change them into less usable forms. Metals dissolved within the oxidation zone may be transported for reprecipitation below the water table. Metals removed from above thus serve to motivate a supergene enrichment within the upper part of the sulfide zone.

Metamorphism is a cyclical or episodic process that has served to change, during the past 4.6 b. y., the form and texture of minerals, including the creation and destruction of economic deposits. Changes in temperature and pressure, or in the quantity of water present, serve to stabilize some minerals, to unstabilize others. Typical results are recombinations and recrystallizations.

Barriers within a conduit tend to influence mineral deposition. These barriers may be of many types. Gougy, impervious, pre-mineral cross-fractures may precipitate cross-breaking seams. Wherever breccia-forming walls grade into gouge-forming walls, the change in the fault-filling coinciding with the contact may form an equally effective barrier and thereby confine the ore-bearing solutions to more competent layers, e. g., when a vein passes from quartzite to shale.

The greatest concentrations of ore are usually just below gently dipping barriers, except that, wherever the solutions have had a marked horizontal component, the ore concentrates on either the hanging-wall or foot-wall side of the barrier. The actual angle deflecting the flow within the barrier determines whether the shoot "bottoms" or "tops" there.

These and other factors relating to the flow of heat and material through the tectonosphere must be included as constraints in any backward extrapolation of the effects of intrusive and extrusive activity on the origin, evolution, and present characteristics of copper deposits.

Other analyses related to the origin, evolution, and present characteristics of the paths of preferential flow within the Earth's tectonosphere include those by Bostrom et al. (1972), Dandurand et al. (1972), Fein (1973), Gill (1973), Griggs (1972), Hackett and Bischoff (1973), Heirtzler (1972), Murray (1970), Nitsan (1973), Roy et al. (1972), Steinhorn (1973), and Zonenshain (1972).

The Relationship between Intrusive and Extrusive Magma and the Low-Velocity Zone as Interpreted by the Tectonospheric Earth Model.

The low-velocity zone is related to intrusive and extrusive activity, according to the Tectonospheric Earth Model, through a common mechanism, i. e., the same driving mechanism motivates (1) magmatic activity and (2) the low-velocity zone, possibly produced by incipient melting in the upper mantle (See, e. g., Lambert and Wyllie, 1970). The postulated incipient melting is caused by the same mechanism that melts the magma that ascends to produce intrusive and extrusive activity. Thus, the low-velocity zone may be considered a complex global network, or system, of horizontal "intrusions" between asthenospheric plates and shells (Tatsch, 1972a: chapter 6). In these regions, the unbalanced force vector on some protomagma is horizontal (i. e., this may occur wherever there are insufficient vertical conduits to convey all the protomagma upward). Magma unable to proceed upward will proceed sideward, if there is a horizontal path, or it will remain in situ, if there are no paths. Intrusive and extrusive activity is closely related, therefore, to both the low-velocity zone and to the paths of preferential flow.

Other studies in related areas have been made by Dimitriyev (1972), Heirtzler (1972), Scheidegger (1972), Shtreys and Tseysler (1972), York and Helmberger (1973), and Zapletal (1968).

The Association between Intrusive and Extrusive Magmatism and Tectonic Activity as Interpreted by the Tectonospheric Earth Model.

Basaltic and rhyolitic magmatism is associated with crustal thinning which, in turn, may be caused by (1) crustal tension, (2) heating from below, or (3) a combination of these. Because the amount of crustal tension and heating in a given area at a given time is dependent upon contributions from the geometrical, mechanical, thermal, and chemical factors of the subsurface environment, it may be said that intrusive and extrusive magmatism forms a temporal spectrum, or continuum, of compositions ranging from ultra-basicity to ultra-acidity. This compositional continuum repeats in a cyclical but aperiodic manner, according to the Tectonospheric Earth Model. This accounts, according to this concept, for the great variety of basic-to-acid rocks that have been emplaced on the surface of the Earth during the past 4.6 billion years (b. y.) (Tatsch, 1972a).

Other analyses in related area include those by Azhgirei (1968), Bain (1968), Benerjee and Ghosh (1972), Bickford and Van Schmus (1973), Borrello (1972), Cepeda (1973), Corliss (1973), Dixon and Pereira (1973), Gilluly (1971, 1972), Harrison (1972), Heirtzler (1972), Hutchinson and Hodder (1972), Kaula (1972), Khain (1972), McElhinny (1972), McGetchin et al. (1973), McKenzie (1972), Miyashiro (1972, 1973), Sillitoe (1972a, 1972b, 1972c, 1973), Spall (1972), Subhotin et al. (1972), Winterer (1973), and Wright and McCurry (1973).

Intrusive and Extrusive Activity Associated with Uplifts and Thrusts as Interpreted by the Tectonospheric Earth Model.

Many mountain ranges are the product primarily of vertical uplift, rather than of horizontal crustal compression as was once thought (See, e. g., Eardley, 1963). The Laramide Rockies, for example, east of the Paleozoic miogeosyncline, were formed primarily by intrusive activity on a large scale consisting essentially of oval or irregularly broad shapes having structural relief of a few hundred meters (e. g., the Bowdoin Dome) to 12 km (e. g., the Wind River Uplift). Later Tertiary faulting modified considerably some of these Laramide "intrusives". Younger Laramides were modified by erosion, sedimentation, and subsequent intrusive and extrusive activity. When the thrust faults of the Laramide province are charted, they prove to be, for the most part, marginal to the uplifts or intrusives. There is general correlation between the amount of uplift and the type and prominence of border thrusts, suggesting a common causal mechanism. This association is very pronounced in cases of large uplifts (i. e., 6 km or more) but they may be almost undetectable in intrusives of low and intermediate uplift.

These and similar relationships between uplifts and thrusts suggest that vertical uplift (or intrusion from below) was the primary deformation and that thrusting was a secondary lateral deformation caused by unbalanced force vectors acting vertically and obliquely downward, i. e., such as would be evidenced by gravity sliding, flowing, and large-scale erosion, for example (Tatsch, 1972a).

Many mineral deposits appear to be associated with (1) the volcanism of intermediate and silicic types related to uplifts and graben faulting; and (2) the flood-basalt extrusions related to massive crustal warping along future rift margins (See, e. g., Baker et al., 1972).

Kimberlites are usually associated with intrusive carbonate (carbonatite), alkali basalts (lamprophyres), or other unusual highly alkalic rocks in regions commonly associated with epeirogenic uplift (Wyllie, 1967, 1969a, 1969b). The genesis of these rocks appears to involve the localization of volatiles or partial melting in the upper mantle. Because this genesis apparently reflects long-lived, deep-seated mantle behavior, the spatio-temporal relationships of kimberlite petrology and mineralogy is of interest to the origin, evolution, and accumulation of other mantle-derived minerals.

Very few proposals exist for the origin of the volatiles that appear to be involved not only in kimberlite emplacement but also in epeirogenic uplift (See, e. g., McGetchin et al., 1973; Tatsch, 1972a).

Related analyses include some by Damon (1972), Faure (1972), Hills (1972), and Trofimov (1972).

Subduction, Obduction, and Other Omniductive Activity as Contributions to the Evolution of Mineral Deposits during the Past 4.6 Billion Years.

Many minerals seem to arrive at the Earth's surface as a result of sea-floor spreading, crustal plate motion, continental drift, and other omniductive movements involving lithospheric plates and the magmas associated therewith. It is well to consider briefly how these geophenomena are interpreted as constituting essential phenomena in the evolution of mineral deposits.

When viewed on a global scale and over a long period of time, sea-floor spreading and other forms of crustal rifting appear to represent the surficial behavior of the Earth in those areas where the Earth's internal behavior causes crustal tension. Crustal plate motion, subsurface plate and block motion, and continental drift, in a similar manner, represent the surficial behavior of the Earth in those areas where the resultant of the Earth's internal behavior is representable as a tangential unbalanced force vector acting on a plate, on a block, or on a continent to produce motion of that block, plate, or continent relative to the bulk of the Earth or to a datum therein (Tatsch, 1972a).

In short, sea-floor spreading, crustal plate motion, subsurface plate and block motion, and continental drift, according to the Tectonospheric Earth Model, are surficial adjustments to compensate for the disequilibrated and ever-changing behavior of the Earth's interior. The Earth's surficial behavior is a temporarily delayed attempt of the Earth's surface to adjust itself to the ceaseless equilibrating behavior of its interior. In some cases a block may be moved to effect equilibration; in others, a plate may be moved. In still other cases, a rift may be formed to effect the equilibration. Sometimes, worldwide mountain ranges grow from these rifts in a single orogeny. At other times, an earthquake occurs when a block or plate adjusts itself to a more nearly equilibrated position in the never-ending attempt of the Earth to assume a condition of minimum energy. These and related matters regarding the Earth's behavior are discussed in greater detail in Tatsch (1972a).

The Evolution of the Mineral Deposits during the Past 4.6 Billion Years according to the Tectonospheric Earth Model.

When the Tectonospheric Earth Model is used to analyze the evolution of the mineral deposits during the past 4.6 b. y., it is convenient, because of the inherent geometrical, mechanical, thermal, and chemical constraints of the model, to consider three aspects of the evolution of the mineral deposits: (1) their most probable evolutionary behavior during the period 4.6 b. y. to 3.6 b. y. ago; (2) their most probable evolutionary behavior during the period 3.6 b. y. to 0.6 b. y. ago; and (3) their most probable evolutionary behavior during the period 0.6 b. y. ago to the present. This is done in the following sections.

The Evolution of the Mineral Deposits during the Period 4.6 to 3.6 Billion Years Ago.

For the purpose of this analysis, it is assumed, for reasons stated above, that the Earth's surface 4.6 b. y. ago was devoid of any surficial features other than those that would have been genetic to an accretionary process of planetary formation from the solar nebula at that time (Tatsch, 1972a: chapter 1). The evolutionary behavior of the mineral deposits during the period 4.6 b. y. to 3.6 b. y. ago would have been motivated by the Earth's internal behavior plus the influence of extraterrestrial factors operating upon the Earth during that period.

According to the Tectonospheric Earth Model, the fragments of the five "terrestrial" octants of Earth Prime became accreted onto a 5400-km primordial Earth during the period 4.6 b. y. to 3.6 b. y.

ago. Present evidence does not define clearly exactly when this accretion occurred within this one-billion-year span. Nor is it necessary to this discussion that this be known. Suffice it here to say merely that the evolution of the mineral deposits during the period 4.6 b. y. to 3.6 b. y. ago depended primarily upon two basic aspects of the Earth's behavior during that period: (1) the mineral "mix" within the accreting fragments of Earth Prime was not as homogeneous as was that of the surface of the 5400-km primordial Earth; and (2) the primordial Earth, as it accreted to its present size, behaved in accordance with the geometrical, mechanical, thermal, and chemical aspects described in Chapters 2 and 3.

It is of interest to recall, in this connection, that many of these accreting fragments bore a marked resemblance to the rocks that now comprise the Moon, the asteroids, and the meteorites (Tatsch, 1972a: chapter 19; 1974a).

The above-described behavior produced the following constraints upon the evolutionary behavior of the mineral deposits during the period 4.6 b. y. to 3.6 b. y. ago: (1) the surface of the 5400-km primordial Earth was sialic material cut by three mutually-orthogonal belts of extruding mafic and ultramafic material; and (2) the accreting material (from the fragments of the five terrestrial octants of Earth Prime) comprised a complete spectrum of material, ranging from (a) fragments of iron, nickel-iron, and other metals that might have been contained within the molten and plastic parts of Earth Prime prior to its fragmentation, to (b) sialic material (from the surface of Earth Prime) that was similar to that of the surface of the 5400-km primordial Earth.

These constraints produced the following evolutionary behavior within the Earth's surface during the period 4.6 b. y. to 3.6 b. y. ago: (1) the initial sialic minerals on the surface of the primordial 5400-km Earth and on the erstwhile surface of Earth Prime were thoroughly metamorphosed by the effects of the accretionary activity and of the geometrical, mechanical, thermal, and chemical behavior of the primordial Earth as it grew from about 5400 km to its present size; and (2) as the size of the primordial Earth grew (at an average rate of about one millimeter per year), the geometrical, mechanical, thermal, and chemical effects of the Earth's internal behavior were felt mainly through the primordial orogenic belts of the Earth during the period 4.6 b. y. to 3.6 b. y. ago, i. e., through the wedge-belts of activity of the Tectonospheric Earth Model.

This evolutionary behavior within the Earth's surface (as it grew at an average rate of about 3 micrometers per day) produced the following effects upon the distribution patterns of the mineral deposits during that period: (1) the igneous, metamorphic, and sedimentary processes associated with the mineral deposits tended to form three mutually-orthogonal belts upon the surface of the Earth; and (2), as the Earth grew, at a rate of a few micrometers per day, it tended to continually equilibrate its condition to one of minimum energy by modifying the upper few micrometers of its surface each day, through the geometrical, mechanical, thermal, and chemical aspects of its driving mechanism (Chapters 2 and 3), thereby redistributing, metamorphosing, and otherwise affecting the mineral deposits to the extent that would be expected as a result of the geometrical, mechanical, thermal, and chemical effects of the basic driving mechanism under these conditions.

The most probable distribution pattern of the mineral deposits at the end of this period, i. e., 3.6 b. y. ago, may be summarized: (1) the youngest mineral belts (i. e., those 3.6 b. y. old on the present time scale) were aligned along various segments of three mutually-orthogonal belts; and (2) the older mineral belts consisted of "fossil" belts that had been overprinted, metamorphosed, and otherwise modified to the extent that would be expected as a result of the geometrical, mechanical, thermal, and chemical effects of the Earth's internal behavior during the period 4.6 b. y. to 3.6 b. y. ago.

The Evolution of the Mineral Deposits during the Period 3.6 to 0.6 Billion Years Ago.

For the purpose of this analysis, it is assumed, for reasons previously stated, that the Earth's surface 3.6 b. y. ago was that which would be expected from the evolutionary behavior described in the previous section. It is assumed further that, because most extra-terrestrially induced effects upon the Earth ceased prior to 3.6 b. y. ago, the Earth's behavior during the past 3.6 b. y. was internally motivated (Tatsch, 1972a). The evolutionary behavior of the mineral deposits during the period 3.6 b. y. to 0.6 b. y. ago would have been motivated almost entirely by the Earth's internal behavior during that period.

This behavior produced the following constraints upon the evolutionary behavior of the mineral deposits during the period 3.6 b. y. to 0.6 b. y. ago: (1) the surface of the Earth 3.6 b. y. ago consisted of three mutually-orthogonal orogenic belts plus modified older "fossils" as described in the previous section; and (2) new activity occurred primarily within the wedge-belts of activity associated with the Earth during the period 3.6 b. y. to 0.6 b. y. ago.

These constraints produced the following evolutionary behavior within the Earth's surface during the period 3.6 b. y. to 0.6 b. y. ago: (1) the younger mineral belts (i. e., those that are now 3.6 b. y. old) were metamorphosed by the effects of the geometrical, mechanical, thermal, and chemical behavior of the Earth during this period; and (2) some of the older "fossil" belts (i. e., those older than 3.6 b. y.) were modified by uplift, erosion, metamorphism, overprinting, and by other effects expected from the geometrical, mechanical, thermal, and chemical behavior of the Earth during that period.

This evolutionary behavior within the Earth's surface produced the following effects upon the distribution patterns of the mineral deposits during the period 3.6 b. y. to 0.6 b. y. ago: (1) the igneous, metamorphic, and sedimentary processes associated with the mineral deposits tended to form three mutually-orthogonal belts upon the surface of the Earth at any time during that period; and (2) the Earth tended to continually equilibrate its condition to one of minimum energy through the geometrical, mechanical, thermal, and chemical aspects of its driving mechanism (Chapters 2 and 3), thereby redistributing, metamorphosing, and otherwise affecting the mineral deposits to the extent that would be expected from the geometrical, mechanical, thermal, and chemical effects of the Earth's basic driving mechanism under these conditions.

The most probable distribution pattern of the mineral deposits at the end of this period, i. e., 0.6 b. y. ago, may be summarized: (1) the youngest mineral belts (i. e., those of the Upper Precambrian) were aligned along various segments of three mutually-orthogonal belts; and (2) the older mineral belts consisted of "fossil" belts that had been overprinted, metamorphosed, and otherwise modified as would be expected to result from the geometrical, mechanical, thermal, and chemical effects of the Earth's internal behavior during the period 3.6 b. y. to 0.6 b. y. ago.

The Evolution of the Mineral Deposits during the Period 0.6 Billion Years Ago to the Present.

For the purpose of this analysis, it is assumed, for reasons previously stated, that the Earth's surface at the end of the Precambrian was that which would be expected from the evolutionary behavior described in the previous section. It is assumed further that the Earth's behavior during the Phanerozoic was basically internally motivated (Tatsch, 1972a). The evolutionary behavior of the mineral deposits during the Phanerozoic have been motivated, under this concept, almost entirely by the Earth's internal behavior during this period.

This behavior produced the following constraints upon the evolutionary behavior of the mineral deposits during the Phanerozoic: (1) the surface of the Earth at the end of the Precambrian consisted of three mutually-orthogonal orogenic belts plus modified older "fossils" as described in the previous section; and (2) new activity has occurred primarily within the wedge-belts of activity associated with the Earth during the Phanerozoic.

These constraints have produced the following evolutionary behavior within the Earth's surface during the Phanerozoic: (1) the younger mineral belts (i. e., those of latest Precambrian age) were metamorphosed by the effects of the geometrical, mechanical, thermal, and chemical behavior of the Earth during the Phanerozoic; and (2) some of the older "fossil" belts (Precambrian) were modified by uplift, erosion, metamorphism, overprinting, and by such other effects that might be expected from the geometrical, mechanical, thermal, and chemical behavior of the Earth during this period.

This evolutionary behavior within the Earth's surface has produced the following effects upon the distribution patterns of the mineral deposits during the Phanerozoic: (1) the igneous, metamorphic, and sedimentary processes associated with the mineral deposits have tended to form three mutually-orthogonal belts upon the surface of the Earth at any time during the Phanerozoic; and (2) the Earth has tended to continually equilibrate its condition to one of minimum energy through the geometrical, mechanical, thermal, and chemical aspects of its driving mechanism (See Chapters 2 and 3), thereby redistributing, metamorphosing, and otherwise affecting the mineral deposits to the extent that would be expected from the geometrical, mechanical, thermal, and chemical effects of the Earth's basic driving mechanism under these conditions.

The most probable distribution pattern of the mineral deposits at the present time, according to this concept, may be summarized: (1) the youngest mineral belts (for example, the Tertiary copper belts and the Tertiary petroleum belts) should be aligned along various segments of three mutually-orthogonal belts; and (2) the older mineral belts consist of "fossil" belts that have been overprinted, metamorphosed, and otherwise modified as would be expected to result from the geometrical, mechanical, thermal, and chemical aspects of the Earth's internal behavior during the Phanerozoic.

Chapter 6

CENOZOIC SEISMOTECTONOMAGMATIC BELTS AND THE ASSOCIATED COPPER DEPOSITS

An understanding of the origin, evolution, and present characteristics of the Cenozoic copper deposits of the world is of interest because such a large portion of the Earth's copper deposits appear to be associated with the seismotectonomagmatic belts that have developed during the Cenozoic. It is well to attempt to define these deposits in terms of the geometrical, mechanical, thermal, and chemical aspects of the Earth's behavior that have produced the deposits during the 4.6 b. y. that the Earth is believed to have been in existence.

There are many approaches that may be used to make the above-proposed analysis. The approach used in this book is to attempt to show how the present copper deposits have originated, have evolved, and have become deposited in their present environs in accordance with a single, long-lived, deep-seated, global mechanism that has been operating within the Earth during the past 4.6 b. y. The proposed mechanism is the Tectonospheric Earth Model. This model has been described in Chapter 2; some of its geometrical, mechanical, thermal, and chemical aspects have been summarized in Chapters 3 and 4. On that basis, it is well to summarize the origin, evolution, and present characteristics of the Earth's surficial features that have contributed to the present distribution of the Earth's copper deposits.

The Origin, Evolution, and Present Characteristics of the Earth's Surficial Features.

Basic to the seismotectonomagmatic-belt concept of the Tectonospheric Earth Model is that the Earth's present surficial features are a consequence of the Earth's internal behavior during the 4.6 b. y. that the Earth is believed to have been in existence. The salient aspects of the correlations between the Earth's surface and its interior have been discussed in Chapter 3. It is well here to look more closely at those aspects of the Earth's internal behavior that have contributed to the origin, evolution and present characteristics of the Earth's copper deposits. To do this, it is necessary to look at both the continents and the oceans.

The continents comprise roughly 1/3 of the Earth's present surficial area. Continental sedimentary thicknesses vary widely throughout the Earth. In a given region, these thicknesses range from an average of a few km to a maximum of about 15 km. Under these sedimentary thicknesses are roughly 20 km of metamorphic rocks (gneisses and granites). Outcrops of these occur through the sedimentary thicknesses in all continents. A deeper basaltic layer lies beneath the crustal metamorphic rocks. A seismic velocity anomaly (the Mohorovicic discontinuity) exists at the top of this basaltic layer. The total crustal thickness ranges from about 25 km to about 75 km.

The oceans, in contrast to the continents, are underlain by a relatively uniform basaltic crust, averaging roughly 10 to 12 km (See, e. g., Jackson et al., 1972; Reyment, 1973; Tatsch, 1972a: chapter 15; 1975d). Unlike the orogenic-cratonic structure of the continents, that of the oceans seems to be simpler and younger (See, e. g., Tatsch, 1972a: chapter 7; 1975d). It is well to summarize some of the salient aspects of the differences that exist between the continents and the oceans:

a. About 8% of the Earth's surface (i. e., roughly 25% of the present continental areas) comprises very old, highly-metamorphosed Precambrian rocks. Rocks of this type are rarely found in the oceans. However, the observed paucity of Precambrian rocks in the oceans could be the result of incomplete observational data for the oceans; and the possibility has not been eliminated that Precambrian (and Paleozoic) sediments may exist today within Layer 3 of the oceanic crust (See, e. g., Brown, 1974b; Rona and Dietz, 1974; Tatsch, 1972a: chapter 15).

b. At various time, the shield areas have undergone gentle tilting and associated submersion along the affected edges.

c. At other times, the Earth's internal behavior has caused large-scale, vertical (epeirogenic) movements that have affected some shield areas, causing more general submersion. Other shields have undergone little or no epeirogenic subsiding.

d. Occasionally, crustal rifting has separated some of the oldest shield areas. Other shields have remained more-or-less unbroken ever since Precambrian times.

An analysis of the origin, evolution, and present characteristics of the Earth's copper deposits must consider the effects of these and related horizontal and vertical movements that the oceans and continents have undergone while the copper deposits were forming.

The Earth's Copper Deposits Associated with the Cenozoic Seismotectonomagmatic Belts: An Overview.

The present distribution patterns of the Earth's copper deposits are identifiable, according to the seismotectonomagmatic-belt concept, with the surficial manifestations of the Earth's internal activity associated with the 12 quadrispherical arcs of that model. The present configurations of these 12 arcs may be listed, together with some of the applicable areas that are known to contain copper deposits:

a. Bengal-to-Kermadecs. Australia, Bangladesh, Brunei, Burma, Indonesia (including West Irian, Java, Kalimantan, and Sumatra), the Khmer Republic, Malaysia, Papua New Guinea, southern Philippines, South Vietnam, Sri Lanka, Thailand, and the associated offshore and sea-bottom areas.

b. Kermadecs-to-Galapagos. None known, but some may exist in association with sea-bottom nodules.

c. Galapagos-to-Gibraltar. Central America, northern South America (primarily Colombia, Venezuela, Trinidad-Tobago, the Guianas, and Surinam), northwestern Africa, Spain, and the associated offshore and sea-bottom areas.

d. Gibraltar-to-Bengal. Spain, northern Africa, southern Eurasia, the Middle East, Iran, Pakistan, India, and the associated offshore and sea-bottom areas.

e. Bengal-to-Aleutians. Indonesia, Malaysia, southeastern Asia, northeastern Asia, and the associated offshore and sea-bottom areas, including the South China Sea, the East China Sea, the Sea of Japan, the Sea of Okhotsk, and the Bering Sea.

f. Aleutians-to-Galapagos. Southern Alaska, western Canada, western United States, western Mexico, Central America, and associated offshore and sea-bottom areas.

g. Galapagos-to-Bouvet. Western South America, eastern Argentina, and the associated offshore and sea-bottom areas.

h. Bouvet-to-Bengal. Southern Africa, southeastern Africa, southern India, southeastern India, and the associated offshore and sea-bottom areas.

i. Kermadecs-to-Aleutians. Southern Alaska, various islands between Alaska and the Tonga-Kermadec complex, and the associated offshore and sea-bottom areas.

j. Aleutians-to-Gibraltar. Western and northern Alaska, Canadian Arctic Islands, western and southern Greenland, western Europe, and the associated offshore and sea-bottom areas.

k. Girbraltar-to-Bouvet. Western Africa and the associated offshore and sea-bottom areas.

l. Bouvet-to-Kermadecs. Antarctica, New Zealand, the Tonga-Kermadec complex, the associated islands, and the associated offshore and sea-bottom areas.

In considering these listings, it is suggested that the reader use a globe with 3 rubber bands to represent the surficial traces of the 12 quadrispherical arcs of the model. When properly placed on the globe, the 3 rubber bands will form 6 orthogonal intersections associated with the 6 points of the model: Aleutians, Galapagos, Gibraltar, Bengal, Kermadecs, and Bouvet (Chapter 2).

It is necessary to recall that the above listings, insofar as they pertain to the origin, evolution, and present characteristics of copper deposits, are based on the first-order outputs of the Tectonospheric Earth Model. These outputs have a computed nominal probable error of 1571 km (Tatsch, 1972a: chapter 6). Resolutions smaller than 1571 km are not intended when making first-order analyses such as this. The resolutions obtainable with this model when using its higher-order outputs are one, two, and three orders better (i. e., 157.1 km, 15.71 km, and 1.571 km), depending upon the amount of time and funding allocated to a given area.

No adjustments have been made in the above listings to correct the first-order predictions of the model for the differential horizontal movements that have occurred, between the Earth's present surficial features and the subtectonospheric fracture system of the model (Tatsch, 1972a: chapter 6), since the time that the copper deposits were emplaced in their present locations.

In the higher-order analyses, these adjustments, amounting to translations of as much as 140 km (and appropriate rotations) for some Cenozoic copper deposits, are made for the regions being analyzed. More-complex adjustments must be made in analyses involving "older" copper deposits. These adjustments may amount to 300 km for Mesozoic deposits (Chapter 7) and 500 km for Paleozoic deposits (Chapter 8).

The important point regarding these adjustments is that the Tectonospheric Earth Model provides a method for determining them to any degree of accuracy, limited only by the amount of time and funding available. Also, because of the "limit stop" aspects of the orogenic-cratonic evolution of the continents (Tatsch, 1972a: chapter 7), the <u>sum</u> of these adjustments rarely exceeds 3500 km, except that adjustments for some of the oldest Archean copper deposits could amount to as much as 4900 km under the seismotectono-magmatic-belt concept of the Tectonospheric Earth Model. No <u>single</u> adjustment under this concept amounts to more than a few km, because individual adjustments are computed and applied incrementally in the entire gamut of the spatio-temporal frame spanning 4.6 b. y. to a depth of 1000 km.

To see how well these concepts apply to the Earth's actual copper deposits, some representative, Cenozoic copper deposits of each continent are analyzed in the following sections.

<u>The Cenozoic Copper Deposits of North America</u>.

The Cenozoic copper deposits of North America, according to the seismotectonomagmatic-belt concept of the Tectonospheric Earth Model, are identifiable with the surficial manifestations of the Earth's internal activity along applicable segments of the Aleutians-to-Galapagos, Galapagos-to-Gibraltar, and Aleutians-to-Gibraltar quadrispherical arcs of the model. To see how these predictions correlate with the facts, it is well to consider some representative, Cenozoic copper deposits of that area:

a. <u>Porphyry Deposits of the United States</u>.

(1). <u>Location</u>. Western United States.

(2). <u>Description</u>. These porphyries range in age from Paleozene to Oligocene. Typical ages may be listed: Esperanza (62 million years), Ajo (62), Ray (62), Inspiration (60), Safford (58), Morenci (58), Bingham (37), and Climax (30).

b. <u>Porphyry Deposits of Canada</u>.

(1). <u>Location</u>. British Columbia.

(2). <u>Description</u>. These porphyries range in age from Paleocene to Miocene. Typical ages may be listed: Ajax, BMC Moly, Newman, Red Bird, Granisle, and Maggie (Paleocene); Bell Copper, Catface and Berg, Gibraltar, and Vancouver (Eocene); Mount Washington (Oligocene); and Quartz Creek (Miocene).

c. <u>Porphyry Deposits of Panama</u>.

(1). <u>Location</u>. Panama.

(2). <u>Description</u>. These porphyries are of the same

age (37 m. y.) as are those of Bingham (Utah) and Vancouver Island (British Columbia). They appear to be related to the Tertiary volcanic rocks. Typical deposits are those at Botija and Petaquilla.

d. Apache Leap Tuff.

(1). Location. Inspiration, Arizona.

(2). Description. These Miocene porous carbonate-bearing rocks were exposed to ground water from weathering copper-sulfide deposits.

e. Matahambre.

(1). Location. Cuba.

(2). Description. These massive chalcopyrite lenses and veins appear to be unrelated to either volcanic or intrusive rocks.

f. Puerto Rico.

(1). Location. Antilles Islands.

(2). Description. These porphyry deposits were formed during a period of intruso-extrusive activity of the early Tertiary. They are overlapped by middle and upper Tertiary coastal-plain carbonate rocks. Ecologically-induced resistance has prevented operation of the two mines that comprise these deposits, i. e., Rio Vive and Takama.

The Cenozoic Copper Deposits of South America.

The Cenozoic copper deposits of South America, according to the seismotectonomagmatic-belt concept of the Tectonospheric Earth Model, are identifiable with the surficial manifestations of the Earth's internal activity along applicable segments of the Galapagos-to-Bouvet and Galapagos-to-Gibraltar quadrispherical arcs of the model. To see how these predictions correlate with the facts, it is well to consider some representative, Cenozoic copper deposits of that area:

a. Antioquia.

(1). Location. Western Colombia.

(2). Description. Unknown quantities of Cu are associated with quartz diorite intrusions of Tertiary age in this part of the Sierra Occidental.

b. Coro-Coro.

(1). Location. Bolivia.

(2). Description. Pliocene sandstones contain these well-known and long-worked deposits. Several cupriferous areas contain disseminated native copper, gypsum, native silver, domeykite, and chalcocite. The copper-bearing beds are much lighter in color than are the prevailing deep-red sandstones.

c. Porphyry Deposits of Peru.

(1). Location. South America.

(2). Description. These porphyry copper deposits range in age from the Paleocene (Cerro Verde and Toquepala) to the Miocene (Michiquillay).

d. Porphyry Deposits of Ecuador.

(1). Location. South America.

(2). Description. The copper deposits of Ecuador are quite young, Chaucha being only about 8 million years old.

e. Porphyry Deposits of Chile.

(1). Location. South America.

(2). Description. These porphyries range from the

Eocene (Mocha and El Salvador); through the Oligocene (Potrerillos and Chuquicamata); and to the end of the Pliocene (El Teniente and Rio Blanco). Exotica, Sagasca, and Huinquinlipa also appear to be derived from nearby porphyries.

f. Porphyry Deposits of Argentina.

(1). Location. South America.

(2). Description. These porphyry mines (Alumbrera and Catamarca) are of Miocene age.

g. Buena Esperanza.

(1). Location. Chile.

(2). Description. Except for the absence of native copper, these deposits resemble the Keweenaw deposits of Michigan. Bornite, chalcocite, and chalcopyrite occur in the amygdaloidal upper parts of the associated andesite flows.

h. El Abra.

(1). Location. Chile.

(2). Description. The oxidized cap of this deposit, which outcrops to the surface, consists of chrysocolla, atacamite, and pseudo-malachite. The primary zone consists of chalcopyrite and bornite.

The Cenozoic Copper Deposits of Africa.

The Cenozoic copper deposits of Africa, according to the seismotectonomagmatic-belt concept of the Tectonospheric Earth Model, are identifiable with the surficial manifestations of the Earth's internal activity along applicable segments of the Gibraltar-to-Bouvet, Gibraltar-to-Bengal, and Bouvet-to-Bengal quadrispherical arcs of the model. To see how these predictions correlate with the facts, it is well to consider some representative, Cenozoic copper deposits of that area:

a. Atlas Mountains.

(1). Location. Northern Africa.

(2). Description. Only scattered deposits of copper are found in the Atlas Mountains and other Cenozoic manifestations of the Gibraltar-to-Bengal seismotectonomagmatic belt in northern Africa. The same may be said for the west and east coasts of Africa.

The Cenozoic Copper Deposits of Western Europe.

The Cenozoic copper deposits of western Europe, according to the seismotectonomagmatic-belt concept of the Tectonospheric Earth Model, are identifiable with the surficial manifestations of the Earth's internal activity along applicable segments of the Aleutians-to-Gibraltar quadrispherical arc of the model. To see how these predictions correlate with the facts, it is well to consider some representative, Cenozoic copper deposits of that area:

a. England.

(1). Location. Western Europe.

(2). Description. Only small scattered copper deposits of Cenozoic age are found in England and related parts of western Europe.

The Cenozoic Copper Deposits of Southern Europe.

The Cenozoic copper deposits of southern Europe, according to the seismotectonomagmatic-belt concept of the Tectonospheric Earth Model, are identifiable with the surficial manifestations of the Earth's internal activity along applicable segments of the Gibraltar-to-Bengal quadrispherical arc of the model. To see how these predictions correlate with the facts, it is well to consider some representative, Cenozoic copper deposits of that area:

a. Porphyry Deposits of Bulgaria.

(1). Location. Southern Europe.

(2). Description. The deposits at Medet, operated since 1965, have made that country self-sufficient in copper.

b. Porphyry Deposits of Yugoslavia.

(1). Location. Southern Europe.

(2). Description. The most important Yugoslavian mine deposit is at Majdanpek (eastern Serbia) near the Bulgarian border. This mine has been in operation since 1962. Newer discoveries include those at Krivelj, near Majaanpek, which comprise large low-grade deposits.

c. Porphyry Deposits of Rumania.

(1). Location. Southern Europe.

(2). Description. Although copper is not economically worked in Rumania, that country has several promising porphyries, including those at Banat.

d. Porphyry Deposits of Armenia.

(1). Location. Soviet Europe.

(2). Description. These deposits supply important quantities of copper from small porphyries at Kadzharan, Kafan, Aigedzor, Dasrakert, Agarak, Alaguez, and other areas. Molybdenum is an important by-product. This belt continues into Iran.

e. Porphyry Deposits in Greece.

(1). Location. Southern Europe.

(2). Description. The most promising Grecian porphyries are in the northern part of Greece, including those at

Skouries.

f. Skouriotissa and Mavrovouni.

(1). Location. Cyprus.

(2). Description. The Cyprus deposits, which occur in the pillow-lava sector of the ophiolite suite, are rich also in other metals. The Mavrovouni ore body, with an estimated weight of 15 million tons, contains 43% Fe, 48% S, 4% Cu, and some Zn, Au, and Ag. Skouriotissa, a large pryitic lens, contains about 2% Cu.

g. Monte Catini.

(1). Location. Western coast of Italy.

(2). Description. Irregular laccolithic stocks of dibase, with some gabbro, break through Eocene marly limestones and siliceous shales. The ore occurs exclusively in the dibase, particularly in its lower part at or near the contact. The ores contain native copper, in crevices and druses, plus some sulfides (chalcocite, bornite, and chalcopyrite predominating), sometimes massive, but partly in large and small, rounded concretions surrounded by clayey, crushed rock and comprising several sulfides in concentric intergrowth. Monte Catini resembles the amygdaloidal flows of the Lake Superior region of North America.

h. Massa Marittima.

(1). Location. Italy

(2). Description. These large Tuscan veins carry chalcopyrite, pyrite, galena, and sphalerite that cut across Eocene limestone and clay shales. The limestone, unlike the shales, is replaced near the vein by pyroxene, epidote, quartz, and sulfides. Mineralization appears to be associated with a Tertiary granite intrusion.

The Cenozoic Copper Deposits of the Middle East.

The Cenozoic copper deposits of the Middle East, according to the seismotectonomagmatic-belt concept of the Tectonospheric Earth Model, are identifiable with the surficial manifestations of the Earth's internal activity along applicable segments of the Gibraltar-to-Bengal quadrispherical arc of the model. To see how these predictions correlate with the facts, it is well to consider some representative, Cenozoic copper deposits of that area:

a. Ergani.

(1). Location. Turkey.

(2). Description. These deposits produce significant quantities of copper.

b. Metal-Rich Red Sea Sediments.

(1). Location. Red Sea.

(2). Description. Iron-rich, brine-saturated sediments containing anomalous amounts of Cu and Zn are found underneath hot brines within the Discovery Deep of the Red Sea. The geology, chemistry, and physical properties of these sediments have been studied (See, e. g., Degens and Ross, 1969). Their copper content averages about 1.3%. The theoretical implications of these deposits is being studied in connection with the origin of sedimentary deposits in other areas. Of particular interest is the tectonic-control implied by the location of these deposits near a crustal rift (See, e. g., Kesler et al., 1975; Vogt, 1975).

c. Porphyry Deposits of Iran.

(1). Location. Middle East.

(2). Description. Many small Iranian deposits in the Alborz Mountains appear to be an extension of the Cu - Mo porphyry belt of Soviet Armenia (to the west) and of the Chagai

District of Pakistan (to the east). The only well known Iranian porphyry deposit is the 450-million-ton, 1% Cu Sar Cheshmeh. Others have been reported in central and southwestern Iran, including Vazd, northwest of Sar Cheshmeh (See, e. g., Waterman and Hamilton, 1975).

The Cenozoic Copper Deposits of Southern Asia.

The Cenozoic copper deposits of southern Asia, according to the seismotectonomagmatic-belt concept of the Tectonospheric Earth Model, are identifiable with the surficial manifestations of the Earth's internal activity along applicable segments of the Gibraltar-to-Bengal quadrispherical arc of the model. To see how these predictions correlate with the facts, it is well to consider some representative, Cenozoic copper deposits of that area:

a. Nankat.

(1). Location. Turkestan.

(2). Description. Deposits of metallic copper and chalcocite have been found in the Tertiary sandstones and gypsiferous marls west of Kokand. But the lack of large copper deposits suggests that little Cenozoic activity has occurred along this part of the Gibraltar-to-Bengal quadrispherical arc of the Tectonospheric Earth Model.

The Cenozoic Copper Deposits of Australasia and the Southwest Pacific.

The Cenozoic copper deposits of Australasia and the southwest Pacific are identifiable, according to the seismotectonomagmatic-belt concept of the Tectonospheric Earth Model, with the surficial manifestations of the Earth's internal activity along applicable segments of the Bengal-to-Kermadecs and Kermadecs-to-Bouvet quadrispherical arcs of the model. To see how these predictions correlate with the facts, it is well to consider some representative, Cenozoic copper deposits of that area:

a. Borneo.

(1). Location. Malaysia.

(2). Description. These young porphyries lie within the Mamut property just inside the Malaysian side of Borneo. Being developed by the Japanese, this property is scheduled to go on stream later this year.

b. Guadalcanal.

(1). Location. Solomon Islands.

(2). Description. These porphyries include several interesting prospects. On-going exploration work is directed toward a more-clearly defined evaluation of these deposits.

c. Manus.

(1). Location. Papua.

(2). Description. These newly-discovered porphyries include some promising areas.

d. New Guinea.

(1). Location. Eastern New Guinea.

(2). Description. These porphyries contain several interesting prospects, such as Frieda River and Ok Tedi.

e. West Irian.

(1). Location. Indonesian New Guinea.

(2). Description. These porphyries contain several interesting prospects, such as Ertsberg.

f. Bougainville.

(1). Location. Panguna, Bougainville, Solomon Islands.

(2). Description. One of the youngest large deposits, these copper porphyries are only about 2 m. y. old.

g. Philippines Porphyries.

(1). Location. Western Pacific.

(2). Description. These young porphyries appear to have originated from (1) the eruption of andesite volcanoes, (2) the emplacement of diorite or differentiation of the magma into more acidic products, (3) multiple intrusion of hypabyssal rocks, and (4) hydrothermal alteration by fluids carrying metallic and other ions. These porphyries are similar in many respects to the Arizona porphyries, but there are some significant differences. The largest Philippine mine is the Biga. Others are Toledo, Santo Tomas, Marinduque (Sipalay), Ino Capayang Bintakay, Dizon, Taysan, Tawi-Tawi, Surigao, and Negros.

h. Fiji Islands.

(1). Location. Southwest Pacific.

(2). Description. These porphyries contain interesting prospects and suggest that copper may be found in other islands along the Kermadecs-to-Galapagos segment of the Cenozoic equatorial seismotectonomagmatic belt.

i. Seafloor Deposits.

(1). Location. Various oceans.

(2). Description. Extensive deposits of copper-bearing ferromanganese nodules appear to be forming on the seafloor. The origin and mode of concentration of the metals in these nodules are not known, but they appear to represent syngenetic deposition. The Cu content of the nodules varies regionally, with the higher concentrations normally on the west sides of basins within about 1000 km of the equator and roughly 2 to 6 km beneath the surface. Typical contents of the nodules are Mn (25%), Cu (1.5%), Ni (1.5%), Co (0.5%), plus other metals in small quantities. No actual Cu minerals have been reported; the Cu ions appear to be adsorbed by the manganese oxides. Some of the deep-sea cores contained small veins of pure Cu. On the crest of the Ninetyeast Ridge in the Indian Ocean, the Glomar Challenger drill penetrated Cu within basalt underneath the sediments, suggesting that hydrothermal deposition had occurred in the bedrock as well as in the sediments. Two sources have been proposed for such deposits: (1) extraction and concentration from the sediments; and (2) derivation from the upwelling mantle material from below.

The Cenozoic Copper Deposits of East Asia.

The Cenozoic copper deposits of East Asia are identifiable, according to the seismotectonomagmatic-belt concept of the Tectonospheric Earth Model, with the surficial manifestations of the Earth's internal behavior associated with the applicable segments of the Aleutians-to-Bengal quadrispherical arc of the model. To see how these predictions of the model correlate with the facts, it is well to consider some representative, Cenozoic copper deposits of that area:

a. Northeastern Islands.

(1). Location. Japan.

(2). Description. The inner zone of these "black ores" comprises a fine-grained mixture of sphalerite, galena, barite, and copper. But most of the copper lies within the Mesozoic and Paleozoic outer zones (Chapters 7 and 8).

The Cenozoic Copper Deposits of the Siberian Platform.

The Cenozoic copper deposits of the Siberian Platform are identifiable, according to the seismotectonomagmatic-belt concept of the Tectonospheric Earth Model, with the Earth's internal activity associated with applicable segments of the Aleutians-to-Bengal quadrispherical arc of the model. To see how these predictions of the model correlate with the facts, it is well to consider some representative, Cenozoic copper deposits of that area:

a. Various.

(1). Location. Siberia.

(2). Description. Only scattered copper deposits of Cenozoic age are found in this region, suggesting that the active portions of this seismotectonomagmatic belt are farther to the east.

Chapter 7

MESOZOIC SEISMOTECTONOMAGMATIC BELTS AND THE ASSOCIATED COPPER DEPOSITS

The Mesozoic copper deposits of the Earth are identifiable, according to the seismotectonomagmatic-belt concept of the Tectonospheric Earth Model, with the surficial manifestations of the Earth's internal activity associated with the 12 quadrispherical arcs of the model. More specifically, according to the model, the origin, evolution, and present characteristics of today's Mesozoic copper deposits are derivable primarily from the Earth's activity within the Mesozoic seismotectonomagmatic belts during that period, as modified during the Cenozoic. The Mesozoic seismotectonomagmatic belts are now "fossil", but they were "current" during that period.

Closer scrutiny of the Tectonospheric Earth Model reveals that the Mesozoic seismotectonomagmatic belts were similar to the Cenozoic (and currently active) belts, in that they formed twelve 10,000-km quadrispherical arcs spanning the 6 points of the subtectonospheric fracture system of the Tectonospheric Earth Model, i. e., the 6 points that now lie roughly 1000 km beneath the Aleutians, Galapagos, Gibraltar, Bengal, Kermadecs, and Bouvet (Tatsch, 1972a: page 144). During the time that has elapsed since the Mesozoic, the crustal plates that contain today's copper deposits have "moved", according to the Tectonospheric Earth Model, with respect to these 6 "fixed" points of the model. Consequently, they have "moved" with respect to the 12 "fixed" quadrispherical arcs that span the 6 points of the model.

It is well to note here that the 6 "fixed" points of the model are constrained to move with the "motions" of the austral and boreal poles of the model (Tatsch, 1972a: chapters 6 and 14). Even in primordial times, the Tectonospheric Earth Model had 4 boreal-to-equatorial quadrispherical arcs and 4 austral-to-equatorial quadrispherical arcs. But the surficial manifestations of the boreal and austral poles of the model have not always been associated with the Aleutians and Bouvet Island (Tatsch, 1972a: chapter 17).

In view of the above constraints, it is difficult to define, in exact terms, the post-Mesozoic "movements" of the crustal plates with respect to the 6 "fixed" points of the model. However, with these reservations in mind, it is meaningful to say that the post-Mesozoic movements of the crustal plates (with respect to any selected datum or reference) have consisted essentially of two types of motion: (1) translation and (2) rotation. The translations have been (a) in a "horizontal" (i. e., spherical-shell) domain and (b) in a "vertical" (i. e., spherical-radius) domain. The rotations, similarly, have been (c) in a "horizontal" (i. e., spherical-shell) domain and (d) in a "vertical" (i. e., central-plane) domain. These 4 types of crustal motion have been discussed elsewhere (Tatsch, 1972a). These types of motion are reviewed briefly here, to the extent that they apply to the origin, evolution, and present characteristics of the Earth's Mesozoic copper deposits.

Of particular interest to copper-deposits analyses are the crustal-plate movements that combine components from two or more of the 4 types of the listed rotatio-translational motion (i. e., a, b, c, and d, above). The geometrical, mechanical, thermal, and chemical aspects of these crustal-plate movements are provided by the three (and four) dimensional omniductive plate motions defined by the Tectonospheric Earth Model (Chapter 4).

Although a discussion of the details of the higher-order

applications of these motions is beyond the scope of this book, some of these have been discussed elsewhere (Tatsch, 1972a: chapters 3, 6, 7, 15, 16, 17, and 18; 1973a). Also, a comprehensive appreciation for the causes and effects of the rotating crustal plates is provided elsewhere (Tatsch, 1972a: figure 17-10-2-3, on page 787, and the accompanying text there). Suffice it here to summarize the aspects of those crustal-plate motions and their causes only insofar as they may have a bearing on the origin, evolution, and present characteristics of today's copper deposits that are associated with the Mesozoic seismotectonomagmatic belts.

Briefly stated, an analysis of the geometrical, mechanical, thermal, and chemical aspects of the Earth's behavior during the post-Mesozoic period, as interpreted by the Tectonospheric Earth Model, suggests why there are certain significant differences between today's Mesozoic and Cenozoic copper deposits. These differences may be observed by considering (1) the origin of today's Mesozoic copper deposits during the Triassic, Jurassic, and Cretaceous; (2) the geometrical, mechanical, thermal, and chemical aspects of the Earth's post-Mesozoic behavior that have modified the Mesozoic crustal plates and have affected, thereby, the evolution of today's Mesozoic copper deposits; and (3) the present characteristics of today's copper deposits as a function of the Earth's post-Mesozoic behavior.

The origin, evolution, and present characteristics of today's Mesozoic copper deposits are associated, as indicated above, with the seismotectonomagmatic belts that developed within the upper 1000 km of the Earth during the Mesozoic (Tatsch, 1972a: page 269 and chapter 7). The Mesozoic seismotectonomagmatic belts are now "fossil", but it is well to take a brief look at how they might have appeared during the early Triassic, as interpreted by the Tectonospheric Earth Model.

Basically, there were, according to the model, twelve 10,000-

km quadrispherical arcs spanning the 6 "fixed" points of the subtectonospheric fracture system, i. e., the 6 points that now lie roughly 1000 km beneath the Aleutians, Galapagos, Gibraltar, Bengal, Kermadecs, and Bouvet, as described earlier. Thus, there were, during the early Triassic, 12 then-current quadrispherical seismotectonomagmatic arcs, similar to today's Quaternary, or current, seismotectonomagmatic arcs. Each of these twelve 10,000-km quadrispherical arcs of the Mesozoic was associated, similarly, with two of the 6 "fixed" points of the subtectonospheric fracture system of the Tectonospheric Earth Model. For example, the 10,000-km quadrispherical arc that is associated today with the eastern seaboard of the United States was associated, during the Mesozoic, with a segment of the equatorial belt that is now identifiable with a great circle passing roughly through the Galapagos, Gibraltar, Bengal, and the Kermadecs (Tatsch, 1972a: page 144). On the surface of the Earth, the Mesozoic, Galapagos-to-Gibraltar quadrispherical arc is still roughly where it was during the Mesozoic. But, with respect to the "fixed" Galapagos-to-Gibraltar arc on the surface of the subtectonosphere, the Mesozoic arc on the surface of the Earth has undergone rotational and translational motions during the time that has elapsed since the early Triassic.

Today's Mesozoic copper deposits originated, according to the Tectonospheric Earth Model, in the subsurficial arcs while the early precursors of these deposits were roughly radially above various segments of the corresponding arcs that span the points on the surface of the subtectonosphere. Thus, today's Mesozoic copper deposits that lie within the eastern seaboard of the United States were formed while the eastern seaboard of that country lay roughly radially above the "fixed" subtectonospheric arc Galapagos-to-Gibraltar.

Similar conclusions may be drawn about today's Mesozoic copper deposits that lie in other parts of the world. To see how well these concepts apply to the Earth's actual Mesozoic copper deposits, some of the applicable deposits from each continent are analyzed in the following sections.

The Mesozoic Copper Deposits of North America.

The Mesozoic copper deposits of North America are identifiable, according to the seismotectonomagmatic-belt concept of the Tectonospheric Earth Model, with the surficial manifestations of the Earth's internal behavior that occurred along segments of the seismotectonomagmatic belts that were "active" in North America during the Mesozoic, as described in the previous section. To see how these predictions correlate with the facts, it is well to consider some representative, Mesozoic copper deposits of that area.

It is not too surprising that some of these lie roughly parallel to some of the Cenozoic copper belts discussed in Chapter 6. Nor is it unexpected, according to the model, that others of these are rotatio-translationally oriented with respect to some of the Cenozoic copper belts. For the sake of brevity, those areas that contain both Cenozoic and Mesozoic copper deposits are included only in Chapter 6.

a. Beatson.

(1). Location. Alaska.

(2). Description. These deposits of Prince William Sound resemble the Cobar deposits of New South Wales. Although now closed, Beatson was once the largest copper producer in Alaska. The ores form elongated replacement bodies along shear zones in graywacke and slate. Main minerals are pyrite, pyrrhotite, chalcopyrite, and cubanite.

b. Bornite-Ruby Creek.

(1). Location. Alaska.

(2). Description. This important copper resource comprises replacement ore bodies in carbonate rocks far from any known igneous source. Numerous lodes of this type, in an 800-km belt along the south slope of Brooks Range, are believed to contain more than 10 million tons of copper metal.

c. Kennecott.

(1). Location. Alaska.

(2). Description. These deposits appear to have originated in nearby volcanic rocks. Flows of Triassic or Carboniferous basalts about 1 km thick are covered by 1/2 km of Triassic dolomitic limestone, which is overlain by a thick series of Jurassic strata. The copper-bearing Nikolai greenstones are amygdaloidal flows of basalt, with specks of native copper, chalcocite, bornite, and chalcopyrite within the amygdules. Copper sulfides occur also in slips, brecciated zones, faults, and even within some of the greenstones themselves.

d. Porphyry Deposits of Canada.

(1). Location. British Columbia.

(2). Description. These porphyries range in age from the Triassic (Highland Valley, Bethlehem Copper, Copper Mountains);through the Jurassic (Lornex, Galore Creek, Stikine Copper, Iron Mask, Lorraine, Island Copper, Brenda, Endako); and through the Cretaceous (Boss Mountain, Casino, Huckleberry, Glacier Gulch).

e. Britanna.

(1). Location. British Columbia.

(2). Description. These deposits occur in a 3 km belt of highly sheared rocks. The associated early Mesozoic slates, tuffs, and intruded dacites and latites now constitute a roof pendant in the Coast Range Batholith. Within the shear zone, the vein-like replacements contain pyrite, chalcopyrite, quartz, serite, and anhydrite.

f. Porphyry Deposits of the United States.

(1). Location. Western United States of America.

(2). Description. These porphyries range in age from the Jurassic (Bisbee) through the Cretaceous (Yerington, Ely, Kingman, Butte, Bagdad, and San Manuel).

g. Santa Rosa Sandstone.

(1). Location. New Mexico.

(2). Description. Similar to Speckled Sandstone Formation of Pakistan except that this is Triassic rather than Permian.

h. Agua Zarca Sandstone.

(1). Location. Nacimiento Range, New Mexico.

(2). Description. These first-cycle sedimentary rocks lie in a restricted basin of the Triassic Chinle Formation along the west flank of the Nacimiento Range.

i. Porphyry Deposits of Mexico.

(1). Location. Northern Mexico.

(2). Description. These porphyries are about 65 m. y. old and lie just south of Arizona. The most prominent mines are Cananea and La Caridad (near Nacozari) in the state of Sonora. These are probably a southern extension of the belt that runs through Bingham Canyon (Utah) and the Arizona mines.

The Mesozoic Copper Deposits of South America.

The Mesozoic copper deposits of South America are identifiable, according to the seismotectonomagmatic-belt concept of the Tectonospheric Earth Model, with the surficial manifestations of the Earth's internal activity that occurred along segments of the seismotectonomagmatic belts that were "active" in South America during the Mesozoic, as described in the initial pages of this chapter. To see how these predictions correlate with the facts, it is well to consider some representative, Mesozoic copper deposits of South America.

It is not too surprising that some of these lie roughly parallel to some of the Cenozoic copper belts discussed in Chapter 6. Nor is it unexpected, according to the model, that others of these are rotatio-translationally oriented with respect to some of the Cenozoic copper belts. For the sake of brevity, those areas that contain both Cenozoic and Mesozoic copper deposits are included only in Chapter 6.

a. Various.

(1). Location. South America.

(2). Description. Only scattered copper deposits of Mesozoic age occur in South America, suggesting a subdued activity of the applicable seismotectonomagmatic belts at that time. Most of the Cu deposition of that area was emplaced during the Jurassic and Cretaceous, suggesting that the applicable seismotectonomagmatic belts of South America became active during the middle of the Mesozoic and that this activity continued during most of the Cenozoic (Chapter 6).

The Mesozoic Copper Deposits of Africa.

The Mesozoic copper deposits of Africa are identifiable, according to the seismotectonomagmatic-belt concept of the Tectonospheric Earth Model, with the surficial manifestations of the Earth's internal behavior that occurred along segments of the seismotectonomagmatic belts that were "active" in Africa during the

Mesozoic, as described in the initial pages of this chapter. To see how these predictions correlate with the facts, it is well to consider some representative Mesozoic copper deposits of Africa.

It is not too surprising that some of these lie roughly parallel to some of the Cenozoic copper belts discussed in Chapter 6. Nor is it unexpected, according to the model, that others of these are rotatio-translationally oriented with respect to some of the Cenozoic copper belts. For the sake of brevity, those areas that contain both Cenozoic and Mesozoic copper deposits are included only in Chapter 6.

a. Macalder.

(1). Location. Kenya.

(2). Description. These deposits comprise gold-copper lenses containing pyrrhotite, chalcopyrite, and pyrite.

b. Insizwa and Pilansberg.

(1). Location. South Africa.

(2). Description. These small magmatic copper sulfide deposits, with some cubanite, resemble the low-grade copper deposits in the Pilansberg Bushveld rocks.

The Mesozoic Copper Deposits of Western Europe.

The Mesozoic copper deposits of western Europe are identifiable, according to the seismotectonomagmatic-belt concept of the Tectonospheric Earth Model, with the surficial manifestations of the Earth's internal behavior that occurred along segments of the seismotectonomagmatic belts that were "active" in western Europe during the Mesozoic, as described in the initial pages of this chapter. To see how these predictions correlate with the facts, it is well to consider some representative, Mesozoic copper deposits of western Europe.

It is not too surprising that some of these lie roughly parallel to some of the Cenozoic copper belts discussed in Chapter 6. Nor is it unexpected, according to the model, that others of these are rotatio-translationally oriented with respect to some of the Cenozoic copper belts. For the sake of brevity, those areas that contain both Cenozoic and Mesozoic copper deposits are included only in Chapter 6.

a. Various.

(1). Location. Europe.

(2). Description. Much of the Triassic of western Europe is copper-bearing. Two of the well-known deposits are listed below.

b. Commern-Mechernich.

(1). Location. Europe.

(2). Description. These low-grade ores near Aix-la-Chapelle, lie under about 40 m of overburden. Chalcopyrite fills the veins within sandstones.

c. Buntsandstein.

(1). Location. Europe.

(2). Description. The Voltzia sandstone near Saarlouis is particularly rich in chalcocite and copper carbonates.

The Mesozoic Copper Deposits of Great Britain.

The Mesozoic copper deposits of Great Britain are identifiable, according to the seismotectonomagmatic-belt concept of the Tectonospheric Earth Model, with the surficial manifestations of the Earth's internal behavior that occurred along segments of the seismotectonomagmatic belts that were "active" in Great Britain during the Mesozoic, as described in the initial pages of this chapter. To see how these predictions correlate with the facts, it is well to consider some representative Mesozoic copper deposits of Great Britain.

It is not too surprising that some of these lie roughly parallel to some of the Cenozoic copper belts discussed in Chapter 6. Nor is it unexpected, according to the model, that others of these are rotatio-translationally oriented with respect to some of the Cenozoic copper belts. For the sake of brevity, those areas that contain both Cenozoic and Mesozoic copper deposits are included only in Chapter 6.

During the late Precambrian and early Paleozoic, the copper belt of Great Britain was part of a geosyncline that extended through the northern parts of North America and Europe. During the Carboniferous, this territory comprised the marginal part of the western European geosyncline. Folding occurred during the Caledonian and Hercynian orogenies. Southern structures of the territory are derived primarily from the Armorican phase of the Hercynian orogeny. Northern structures resulted primarily from the Caledonian folding. Final structural changes occurred during the Neogene as consequences of the Alpine orogeny. Extensive volcanism occurred during the Paleozoic.

During the Lower and Middle Paleozoic, Great Britain was inundated by seas. During most of the Carboniferous (except the lowest Lower Carboniferous), thick coal beds formed in the Midlands and in parts of Scotland. During the Permian, regression occurred and the evolving land areas were covered with lagoons and evaporite beds similar to those of northern Europe. This extended into the Triassic, to be followed by marine periods during the Jurassic and Cretaceous. Beginning with the eogene, Great Britain has been dry.

It is within this framework that the copper deposits of Great Britain developed. One of these is described below.

a. Manchester.

(1). Location. England.

(2). Description. At Alderley Edge and Mottram St. Andrews, south of Manchester, copper ores have been mined from Triassic sandstones and conglomerates, mainly as carbonates and vanadates of copper. These deposits appear to be associated with the same seismotectonomagmatic belt that is associated with the Triassic of western Europe.

The Mesozoic Copper Deposits of Northeast Asia.

The Mesozoic copper deposits of Siberia are identifiable, according to the seismotectonomagmatic-belt concept of the Tectonospheric Earth Model, with the Earth's internal behavior that occurred along segments of the seismotectonomagmatic belts that were "active" in northeast Asia during the Mesozoic, as described in the initial pages of this chapter. To see how these predictions correlate with the facts, it is well to consider some representative Mesozoic copper deposits of northeast Asia.

It is not too surprising that some of these lie roughly parallel to some of the Cenozoic copper belts described in Chapter 6. Nor is it unexpected, according to the model, that others of these are rotatio-translationally oriented with respect to some of the Cenozoic copper belts. For the sake of brevity, those areas that contain both Cenozoic and Mesozoic copper deposits are included only in Chapter 6.

a. Porphyry Deposits of Mongolia.

(1). Location. Erdeinetin, Mongolia.

(2). Description. Not much is known about this recently discovered porphyry deposit.

b. Koryak Peninsula.

(1). Location. Soviet Far East, U. S. S. R.

(2). Description. Unknown quantities of copper lie within the Mesozoic fold belts associated with cordilleran-type structures of this peninsula.

The Mesozoic Copper Deposits of Australasia and the Southwest Pacific.

The Mesozoic copper deposits of Australasia and the southwest Pacific are identifiable, according to the seismotectonomagmatic-belt concept of the Tectonospheric Earth Model, with the surficial manifestations of the Earth's internal behavior that occurred along segments of the seismotectonomagmatic belts that were "active" in Australasia and the southwest Pacific during the Mesozoic, as described in the initial pages of this chapter. To see how these predictions correlate with the facts, it is well to consider some representative, Mesozoic copper deposits of Australasia and the southwest Pacific.

It is not too surprising that some of these lie roughly parallel to some of the Cenozoic copper belts described in Chapter 6. Nor is it unexpected, according to the model, that others of these are rotatio-translationally oriented with respect to some of the Cenozoic copper belts. For the sake of brevity, those areas that contain both Cenozoic and Mesozoic copper deposits are included only in Chapter 6.

a. Cobar.

(1). Location. New South Wales.

(2). Description. These lodes lie in an eroded desert range of older Paleozoic sediments. The ore carries chalcopyrite, magnetite, and pyrrhotite in large lenses. Strong enrichment occurs in the upper levels as oxidized ores. Just below the water table, secondary sulfides occur as chalcocite and chalcopyrite.

Chapter 8

PALEOZOIC SEISMOTECTONOMAGMATIC BELTS AND THE ASSOCIATED COPPER DEPOSITS

The Earth's copper deposits are identifiable, according to the seismotectonomagmatic-belt concept of the Tectonospheric Earth Model, with the surficial manifestations of the Earth's internal activity associated with the 12 quadrispherical arcs of the model. More specifically, according to the model, the origin, evolution, and present characteristics of today's Paleozoic copper deposits are derivable primarily from the Earth's activity within the Paleozoic seismotectonomagmatic belts during that period, as modified during the Mesozoic and Cenozoic. The Paleozoic seismotectonomagmatic belts are now "fossil"; but they were "current", or active, during that period.

Closer scrutiny of the Tectonospheric Earth Model reveals that the Paleozoic seismotectonomagmatic belts were similar to the earlier-described belts of the Mesozoic and Cenozoic, in that they formed twelve 10,000-km quadrispherical arcs spanning the 6 "basic points" of the subtectonospheric fracture system of the Tectonospheric Earth Model, i. e., the 6 points that now lie roughly 1000 km beneath the Aleutians, Galapagos, Gibraltar, Bengal, Kermadecs, and Bouvet (Tatsch, 1972a: page 144). During the time that has elapsed since the Paleozoic, the crustal plates that contain today's Paleozoic copper deposits have "moved" repeatedly, according to the Tectonospheric Earth Model, with respect to these 6 "fixed" points of the model. Consequently, they have "moved" also with respect to the 12 "fixed" quadrispherical arcs that span the 6 points.

It is well to note here that the "fixed" basic points of the model are constrained to move with the "motions" of the austral and boreal poles of the model (Tatsch, 1972a: chapters 6 and 14). Even in primordial times, the Tectonospheric Earth Model had 4 boreal-to-equatorial quadrispheric arcs and 4 austral-to-equatorial quadrispheric arcs. But the surficial manifestations of the boreal and austral poles have not always been associated with the Aleutians and Bouvet areas as they are today (Tatsch, 1972a: chapter 17).

In view of the above, it is difficult to define, in exact terms, the post-Paleozoic "movements" of the crustal plates with respect to the 6 "fixed" points of the model. However, with these reservations clearly in mind, it is meaningful to say that the post-Paleozoic movements of the crustal plates (with respect to any selected datum or reference) have consisted essentially of two types of motion: (1) translation and (2) rotation. The translations have been (a) in a "horizontal" (i. e., spherical-shell) domain and (b) in a "vertical" (i. e., spherical-radius) domain. The rotations, similarly, have been (c) in a "horizontal" (i. e., spherical-shell) domain and (d) in a "vertical" (i. e., central-plane) domain. These 4 types of crustal motion, previously discussed elsewhere (Tatsch, 1972a), may be reviewed briefly here.

Of particular interest to analyses of copper deposits are the crustal-plate movements that combine components from two or more of the 4 types of the listed rotatio-translational motion (i. e., a, b, c, and d, above). The geometrical, mechanical, thermal, and chemical aspects of these crustal-plate movements are provided by the three (and four) dimensional omniductive plate motions defined by the Tectonospheric Earth Model (Chapter 4).

Although a discussion of the details of the higher-order applications of these motions is beyond the scope of this book, some of these details have been discussed elsewhere (Tatsch, 1972a:

chapter 3, 6, 7, 15, 16, 17, and 18; 1973a). Also, a comprehensive appreciation for the causes and effects of the rotating crustal plates is provided elsewhere (Tatsch, 1972a: figure 17-10-2-3, on page 787, and the accompanying text there). Suffice it here to summarize the aspects of the crustal-plate motions and their causes that have a bearing on the origin, evolution, and present characteristics of today's Paleozoic copper deposits.

Briefly stated, an analysis of the geometrical, mechanical, thermal, and chemical aspects of the Earth's behavior during the post-Paleozoic period, as interpreted by the Tectonospheric Earth Model, suggests that there are certain differences between today's Paleozoic and Cenozoic copper deposits. These differences may be observed by considering (1) the origin of today's Paleozoic deposits during the roughly 350 m. y. that span the Paleozoic; (2) the geometrical, mechanical, thermal, and chemical aspects of the Earth's post-Paleozoic behavior that have modified the Paleozoic crustal plates and have affected, thereby, the evolution of today's Paleozoic copper deposits; and (3) the present characteristics of today's Paleozoic copper deposits as a function of the Earth's post-Paleozoic behavior.

The origin, evolution, and present characteristics of today's Paleozoic copper deposits are associated, as indicated above, with the seismotectonomagmatic belts that developed within the upper 1000 km of the Earth during the Paleozoic (Tatsch, 1972a: page 269 and chapter 7). Data bearing on this period of the Earth's history have been analyzed elsewhere (See, e. g., Berg and Sutton, 1975; Brock, 1973; Burke and Dewey, 1973; Creer, 1973; Dalziel et al., 1973; Deutsch and Murthy, 1973; Fahrig, 1971; Goodwin, 1973; Hurley, 1973; Johnson, 1973; Kisvarsanyi, 1974; Lowry, 1974; Muehlberger and Ritchie, 1975; Palmer, 1974; Piper, 1973a; Shackleton, 1973; and Sutton, 1973). The results in many of these analyses suggest that the Paleozoic behavior of the Earth resulted from stress systems that began in the Precambrian (See, e. g., Lowry, 1974).

Paleozoic seismotectonomagmatic belts are now "fossil", but it is well to take a brief look at how they might have looked during the early Cambrian. Basically, there were then, according to the model, twelve 10,000-km quadrispherical arcs spanning the 6 "fixed" basic points of the subtectonospheric fracture system, i. e., the 6 points that now lie roughly 1000 km beneath the Aleutians, Galapagos, Gibraltar, Bengal, Kermadecs, and Bouvet, as described earlier. Thus, there were, during the early Cambrian, 12 then-current quadrispherical seismotectonomagmatic arcs, similar to today's Quaternary, or now-current, seismotectonomagmatic arcs. Each of the twelve 10,000-km quadrispherical arcs of the Paleozoic was associated, similarly, with two of the 6 "fixed" basic points of the subtectonospherical fracture system of the Tectonospheric Earth Model. For example, the 10,000-km quadrispherical arc that is associated today with the Paleozoic copper deposits of the eastern parts of the United States was associated, during the Paleozoic also, with segments of the equatorial seismotectonomagmatic belt that is now identifiable with a great circle passing roughly through the Galapagos, Gibraltar, Bengal, and Kermadecs (Tatsch, 1972a: page 144).

On the surface <u>of the Earth</u>, the Paleozoic quadrispherical arc is still roughly where it was during the Paleozoic. But, with respect to the "fixed" equatorial belt (roughly 1000 km beneath the Galapagos, Gibraltar, Bengal, and Kermadecs) on the surface <u>of the subtectonosphere</u>, the Paleozoic arc on the surface of the Earth has undergone rotational and translational motions during the time that has elapsed since the Cambrian.

Today's Paleozoic copper deposits, according to the Tectonospheric Earth Model, originated in the subsurficial arcs while the substances genetic to these deposits were roughly radially above various segments of the corresponding arcs that span the 6 "fixed" basic points on the surface of the subtectonosphere. Thus, today's Paleozoic copper deposits that lie within the Appalachians were formed while that part of North America lay roughly radially above the "fixed" equatorial seismotecteonmagmatic belt of the model that is now identifiable with the Galapagos, Gibraltar, Bengal, and Kermadecs. In a similar manner, the Paleozoic copper deposits of the Canadian Arctic, and possibly those of Greenland, are identifiable with a Paleozoic boreal-to-equatorial seismotectonomagmatic belt analogous to the present Aleutians-to-Gibraltar seismotectonomagmatic belt.

Similar conclusions may be drawn regarding today's Paleozoic copper deposits that lie in other parts of the world. To see how well these concepts apply to the Earth's actual Paleozoic copper deposits, some representative, Paleozoic copper deposits of each continent are analyzed in the following sections.

The Paleozoic Copper Deposits of Western North America.

The Paleozoic copper deposits of western North America are identifiable, according to the seismotectonomagmatic-belt concept of the Tectonospheric Earth Model, with the surficial manifestations of the Earth's internal behavior that occurred along segments of the seismotectonomagmatic belts that were "active" in western North America during the Paleozoic, as described in the first few pages of this chapter. To see how these predictions correlate with the facts, it is well to consider some representative, Paleozoic copper deposits of that area.

It is not too surprising that some of these lie roughly parallel to some of the Mesozoic copper belts described in Chapter 7. Nor is it unexpected, according to the model, that others of these are rotatio-translationally oriented with respect to some of the Mesozoic copper belts.

For the sake of brevity, those areas that contain Paleozoic

copper deposits, as well as later deposits, are included only in the appropriate portions of Chapter 6 (Cenozoic) and/or Chapter 7 (Mesozoic).

a. Red Beds.

(1). Location. North America.

(2). Description. Copper ores are widely distributed in the "red beds" of much of the Paleozoic and lower Mesozoic of North America. These occur in the arkosic sandstones, conglomerates, and clay shales of Texas, Oklahoma, New Mexico, Arizona, Colorado, Wyoming, Utah, Idaho, and Nova Scotia. These deposits appear to be epigenetic replacements associated with coal, carbonaceous shale, and calcareous or kaolinic sandstone concretions. The occurrences are mainly on the flanks of older continental areas containing Precambrian copper deposits. Some of the individual deposits are described in the following sections.

b. Southwestern U. S. Red Beds.

(1). Location. Southwestern United States.

(2). Description. Numerous uneconomical copper deposits are associated with the late Paleozoic or early Mesozoic red sandstones of the southwestern United States. Host-rock strata, interbedded with the red strata, are mostly grey to weathered yellow in color. Nonmarine feldspathic or arkositic carbonaceous sandstones are the most common host rocks. Some of these deposits are reminiscent of uranium and vanadium sandstone deposits, and some transitional formations contain all three metals. The Nacimiento copper deposit near Cuba, New Mexico, contains chalcocite, bornite, covellite, and secondary Cu minerals that replace the fossil wood and form halos around the wood. Clearly epigenetic, the Nacimiento deposit appears to predate the major structural features of that area. Other New Mexico mines include the San Pedro. Arizona mines include those at Bisbee, Clifton, and Silver Bell. Similar to the deposits at Bisbee are those at Cananea, Mexico.

c. Bingham.

(1). Location. Utah.

(2). Description. Near Salt Lake City, these deposits comprise a depth of about 3 km of Carboniferous quartzite with numerous lenses of limestone, all folded into an open syncline. Into this has been injected an irregular spreading mass of quartz monzonite porphyry occupying about 8 km^2. The metallization, which occurred in and near the intrusives, formed a great variety of ore deposits typical of a slowly-cooling intrusive. Some of the limestone lenses near the porphyry have been extensively contact-metamorphosed. Chalcopyrite is the main Cu mineral.

d. San Andres Limestone.

(1). Location. New Mexico.

(2). Description. Like Speckled Sandstone Formation of Pakistan.

e. Mountain City.

(1). Location. Nevada.

(2). Description. These deposits appear to have originated in nearby metavolcanic rocks of the Nelson Formation of the late Mississippian.

f. Meade Peak Phosphatic Shale.

(1). Location. Western Wyoming.

(2). Description. Unknown quantities of Cu are contained in this continental-margin marine black shale formation of the Permian.

g. Shasta County.

(1). Location. California.

(2). Description. These Devonian and Carboniferous deposits contain intrusions of highly siliceous and sodic alaskite porphyry, which is the country rock of almost all of the important copper ores. The copper appears to have become concentrated during the interval between two epochs of intrusion. The ore bodies are large, irregular tabular masses of pyrite and chalcopyrite.

h. White River.

(1). Location. Alaska.

(2). Description. The copper-bearing basaltic amygdaloids comprising these deposits are interbedded with Carboniferous sediments. Placer copper is common in the creeks and some large masses of native copper have been found. These deposits resemble those within basaltic flows from other parts of the world.

The Paleozoic Copper Deposits of Eastern North America.

The Paleozoic copper deposits of eastern North America are identifiable, according to the seismotectonomagmatic-belt concept of the Tectonospheric Earth Model, with the surficial manifestations of the Earth's internal behavior that occurred along segments of the seismotectonomagmatic belts that were "active" in that area during the Paleozoic, as described in the first few pages of this chapter. To see how these predictions correlate with the facts, it is well to consider some of the Paleozoic copper deposits of that area.

It is not too surprising that some of these lie roughly parallel to some of the Mesozoic copper belts described in Chapter 7. Nor is it unexpected, according to the model, that others of these are rotatio-translationally oriented with respect to some of the

Mesozoic copper belts.

For the sake of brevity, those areas that contain Paleozoic copper deposits, as well as later deposits, are included only in the appropriate portions of Chapter 6 (Cenozoic) and/or Chapter 7 (Mesozoic).

The area discussed in this section includes the northeastern United States and the adjacent areas of eastern Canada, i. e., the area bordering on the marginal trough of the folded Appalachians to the east. Some of the included tectonic districts are those of Greenland, New Brunswick, New York, Pennsylvania, Ohio, Kentucky, West Virginia, Indiana, Tennessee, Michigan, Illinois, Mississippi, and Alabama.

a. Bathurst-New Brunswick.

(1). Location. Canada.

(2). Description. These Paleozoic deposits apparently formed at the margin of geosynclines in association with andesitic and dacitic volcanics.

b. Rouyn District.

(1). Location. Northern Quebec, Canada.

(2). Description. Dominated by the large Noranda mine, these replacement deposits lie in dacite breccias or tuffs probably formed by hydrothermal replacement. Generally similar but smaller deposits occur nearby at Waite-Amulet, Normetal, and Aldermas mines.

c. Blue Hill.

(1). Location. Maine.

(2). Description. In this deposit, the Cambrian Ellsworth schists have been intruded by granite. Near the contact, irregular and tabular bodies of pyrite and chalcopyrite have developed in the chlorite-sericite schist. Considerable quantities of quartz and cordierite accompany the ores. Although these are contact metamorphic like many of the deposits of the Cordillera of North America, it is significant that no deposits such as these cordierite-chalcopyrite ores occur in the western part of North America. This probably is attributable to the high temperatures and pressures required to produce cordierite deposits.

d. Milan.

(1). Location. New Hampshire.

(2). Description. These pyritic deposits, with chalcopyrite, form lenses or bed-like masses in Paleozoic schists. The ores appear to be later than the schistosity, having replaced the schist during Paleozoic intrusions. Cordierite and quartz are among the gangue minerals, but neither they nor the sulfides predate the metamorphism of the surrounding schists. Similar deposits occur in nearby portions of the Paleozoic seismotectonomagmatic belt, including those at Ely, Vermont, and at Eustis, Quebec.

e. Franklin Furnace.

(1). Location. New Jersey.

(2). Description. The deposits of northern New Jersey are very complex. The two deposits at Mine Hill and Sterling Hill, about 5 km apart, lie along a belt of Precambrian crystalline limestone adjoined on the west by coarse gneisses of igneous origin. Cambrian limestone covers these deposits to the east and west. Both deposits form canoe-shaped beds in the Precambrian limestone. More than 100 minerals comprise the deposit, probably products of secondary metamorphism. Many veins cut the deposits. The main Cu minerals are chalcopyrite and bornite. Participation in the general deep metamorphism that affected this region during the Precambrian accounts for much of the complexity of the region.

These deposits resemble those of Langban, Sweden.

f. Ducktown.

(1). Location. Tennessee.

(2). Description. These deposits are massive sulfide replacement lenses in folded and metamorphosed Cambrian schists and graywackes. The primary Cu ore consists of chalcopyrite with some bornite. The deposits are elongated, roughly tabular masses striking northeast and mostly dipping southeast. A thin but rich chalcocite zone, enriched by surface waters, lies at a depth of about 15 m. Genesis appears to have included: (1) development of lime silicates and pyrite; (2) crushing by extended dynamometamorphism; and (3) introduction of pyrrhotite and, later, sulfides of Cu, Zn, and Pb. The exact age of this orebody is not known (See, e. g., Dallmeyer, 1975).

The Paleozoic Copper Deposits of Central North America.

The Paleozoic copper deposits of central North America are identifiable, according to the seismotectonomagmatic-belt concept of the Tectonospheric Earth Model, with the surficial manifestations of the Earth's internal behavior that occurred along segments of the seismotectonomagmatic belts that were "active" in that area during the Paleozoic, as described in the first few pages of this chapter. To see how these predictions correlate with the facts, it is well to consider some representative, Paleozoic copper deposits of that area.

It is not too surprising that some of these lie roughly parallel to some of the Mesozoic copper belts described in Chapter 7. Nor is it unexpected, according to the model, that others of these are rotatio-translationally oriented with respect to some of the Mesozoic copper belts.

For the sake of brevity, those areas that contain Paleozoic

copper deposits, as well as later deposits, are included only in the appropriate portions of Chapter 6 (Cenozoic) and/or Chapter 7 (Mesozoic).

The area discussed in this section includes the states of Kansas, Oklahoma, Texas, New Mexico, Nebraska, Iowa, Colorado, Missouri, and Arkansas.

a. Flowerpot Permian Shale.

(1). Location. Southern Kansas and western Oklahoma.

(2). Description. This comprises a second type of Cu deposit generally associated with red beds, i. e., disseminated copper sulfide minerals within thin beds of gray shale. These shales occur within a sequence that includes red sandstone, dolomite, gypsum, and shale. They are the facies equivalents of saline and other evaporite deposits of western Kansas, western Texas (Midland and Delaware basins), and New Mexico. Like the Kupferschiefer of Europe, these deposits have lateral persistence within thin strata, thereby suggesting a syngenetic origin.

b. Lead-Zinc Replacement.

(1). Location. Missouri.

(2). Description. Copper is produced as a by-product of Pb-Zn replacement bodies in carbonate rocks without any apparent igneous affiliation.

c. Flin Flon District.

(1). Location. Manitoba-Saskatchewan, Canada.

(2). Description. These deposits lie in altered Keewatin basic lava flows, tuffs, and coarser fragmental beds intruded by hornblende, lamprophyre, and younger quartz porphyry. The ore body is a large, solid, sulfide lens. The core is massive sulfide, surrounded by disseminated sulfides.

The Paleozoic Copper Deposits of Africa.

The Paleozoic copper deposits of Africa are identifiable, according to the seismotectonomagmatic-belt concept of the Tectonospheric Earth Model, with the surficial manifestations of the Earth's internal behavior that occurred along segments of the seismotectonomagmatic belts that were "active" in that area during the Paleozoic, as described in the first few pages of this chapter. To see how these predictions correlate with the facts, it is well to consider some representative, Paleozoic copper deposits of that area.

It is not too surprising that some of these lie roughly parallel to some of the Mesozoic copper belts described in Chapter 7. Nor is it unexpected, according to the model, that others of these are rotatio-translationally oriented with respect to some of the Mesozoic copper belts.

For the sake of brevity, those areas that contain Paleozoic copper deposits, as well as later deposits, are included only in appropriate portions of Chapter 6 (Cenozoic) and/or Chapter 7 (Mesozoic).

a. Tsumb.

(1). Location. Southwest Africa.

(2). Description. Two hydrothermal replacement lenses in dolomite form the basis for these massive aggregates of chalcocite with related minerals. Limited oxidation has produced many oxidized minerals.

b. Ansirabe.

(1). Location. Madagascar.

(2). Description. Significant Paleozoic mineral

deposits are found at the intersection of the Mahanoro-Tsiribihina fracture zone with the crosscutting axis of magmatic reactivation (Kutina, 1975). The copper deposits extend along the axis of magmatic reactivation and represent the most striking structural control of ore deposition in Madagascar. Both east and west of the axis of reactivation, other deposits are found along equally spaced lines parallel to the axis of magmatic reactivation. The equal spacing of these lines suggests a set of north-northwest-trending, deep-seated fractures or zones of weakness, such as would be expected within a seismotectonomagmatic belt.

c. Namaqualand.

(1). Location. Southwest Africa.

(2). Description. The O'okiep and East O'okiep mines contain low grade copper sulfides disseminated in norite. The origin is uncertain; both magmatic and replacement mechanisms have been proposed.

The Paleozoic Copper Deposits of Europe.

The Paleozoic copper deposits of Europe are identifiable, according to the seismotectonomagmatic-belt concept of the Tectonospheric Earth Model, with the surficial manifestations of the Earth's internal behavior that occurred along segments of the seismotectonomagmatic belts that were "active" in that area during the Paleozoic, as described in the first few pages of this chapter. To see how these predictions correlate with the facts, it is well to consider some of the Paleozoic copper deposits of that area.

It is not too surprising that some of these lie roughly parallel to some of the Mesozoic copper belts described in Chapter 7. Nor is it unexpected, according to the model, that others of these are rotatio-translationally oriented with respect to some of the Mesozoic copper belts.

For the sake of brevity, those areas that contain Paleozoic copper deposits, as well as later deposits, are included only in the appropriate portions of Chapter 6 (Cenozoic) and/or Chapter 7 (Mesozoic).

a. Roras-Vigsnas-Sulitjelma.

(1). Location. Norway.

(2). Description. These pyritic deposits occur in metamorphic schists, including clay slate, chloritic schist, amphibolite, and gabbro intrusions. The ores comprise marly pyrite, chalcopyrite, sphalerite, and pyrrhotite. Flat ore lenses are found along the schistosity or crosscutting it. Roras extends 2 km along its dip.

b. Falun and Bersbo.

(1). Location. Sweden.

(2). Description. These deposits lie in cordierite gneiss. The ores resemble contact-metamorphosed schist and contain cordierite, spinel, biotite, hornblende, and garnet. The mineral succession is magnetite, pyrite, pyrrhotite, sphalerite, and chalcopyrite. Mineralization was accompanied by high temperatures.

c. Orijarvi.

(1). Location. Finland.

(2). Description. Resembling high-temperature cordierite deposits in other parts of the world, these pyritic copper ores appear to have been formed by replacement in a series of limestones and leptites.

d. Zechstein Shales.

(1). Location. Lower Silesia, Poland.

(2). Description. Like Kupferschiefer of Mansfeld,

German Democratic Republic.

e. Russian Permian.

(1). Location. West of Urals.

(2). Description. The Permian sandstones, marls, and conglomerates, extending far to the west of the Urals, contain low-grade chalcocite, chalcopyrite, and other copper minerals.

f. The Kupferschiefer.

(1). Location. Mansfeld, German Democratic Republic.

(2). Description. The Permian Kupferschiefer is a thin-bedded bituminous marine marl or shale roughly half a meter thick. It contains Cu, Pb, Zn, and Ag in economic concentrations. The equivalent Zechstein shales of Lower Silesia, Poland, contain economic concentrations of Cu in some areas and of Pb in others. Many Europeans consider the mineralization of the Kupferschiefer to be syngenetic and to be related to formations that are interpretable in terms of lithologic facies. Most advocates of the syngenetic hypothesis also suggest some extraordinary way of concentrating greater than usual amounts of Cu from sea water. Submarine mineral springs and erosion of older Cu ores are among the sources suggested. Other concepts propose: (1) heavy metals concentrated in black shales and phosphorites by upwelling nutrient-rich waters; (2) an epigenetic episode of mineralization by saline ground waters derived from the Zechstein salt deposits overlying the Kupferschiefer (Peryt, 1975); (3) metal-rich thermal ground waters such as those found in wells near the Salton Sea; and (4) the soluble trace metals in the Lower Permian red sandstones over which the Kupferschiefer sea transgressed (Renfro, 1974). The Kupferschiefer may contain several tens of millions of tons of rock averaging 0.3% Cu.

g. Lubin.

(1). Location. Western Poland.

(2). Description. These "white-bed" Weissliegendes deposits appear to be related to the "red-bed" Rotliegendes deposits, such as the Kupferschiefer, which overlie them in many regions. But their exact origin and evolution are not known (See, e. g., Peryt, 1975).

h. Rammelsberg.

(1). Location. Northern slope of Harz Mountains, Germany.

(2). Description. Near Goslar, these deposits are enclosed in Devonian slates forming an overturned anticline dipping northward. The slates have been considerably deformed. The main Cu ores are chalcopyrite and related sulfides within a fissure vein more-or-less conformable with the enclosing slates.

i. Rio Tinto.

(1). Location. Huelva, Spain.

(2). Description. Once known for their gold content, these deposits that now produce copper and sulfur comprise the largest pyritic copper deposits of the world. Associated with sheared quartz porphyry intrusives in Paleozoic slates, these deposits are large lenses of solid sulfide surrounded by disseminated sulfides and sericitized wall rock. The ores carry chalcocite in a shallow supergene zone. Controversial in origin, these ores could have had either a magmatic or a hydrothermal history, with most evidence probably favoring the latter.

The Paleozoic Copper Deposits of Asia.

The Paleozoic copper deposits of Asia are identifiable, according to the seismotectonomagmatic-belt concept of the Tectonospheric Earth Model, with the surficial manifestations of the Earth's internal behavior that occurred along segments of the seismotectonomagmatic belts that were "active" in that area during the Paleozoic, as described in the first few pages of this chapter. To see how these predictions correlate with the facts, it is well to consider some of the Paleozoic copper deposits of that area.

It is not too surprising that some of these lie roughly parallel to some of the Mesozoic copper belts described in Chapter 7. Nor is it unexpected, according to the model, that others of these are rotatio-translationally oriented with respect to some of the Mesozoic copper belts.

For the sake of brevity, those areas that contain Paleozoic copper deposits, as well as later deposits, are included only in the appropriate portions of Chapter 6 (Cenozoic) and/or Chapter 7 (Mesozoic).

a. Ural District.

(1). Location. U. S. S. R.

(2). Description. Important mines within the Urals include those at Kabanski, Volkhovski, Turinski, Levikhinski, Degtiarski, Pyshminski, Karabahski, Uchalinski, Sibaski, Gaiski, Buribaiski, and Bliavinski. These are largely native Cu and Cu-Zn. They are accompanied by Au, Ag, and pyrite that sometimes comprises 80% of the ore. The principal ore bodies are concentrated in the central and southern Urals. Those located to the west of the mountains are normally oxidized, while those to the east are richer in sulfides. The copper content is roughly twice that of porphyries.

b. Timna.

(1). Location. Elat, Israel.

(2). Description. This stratiform copper deposit is of lowermost Cambrian age. The copper-bearing strata consist of fine-grained, bedded sandstones, shales, clays, and lenses, plus layers of sandy dolomite suggesting emplacement in a shallow sea. They overlie a Precambrian basement complex which contains some lead and copper sulfides. The deposits lie along the fringes of the Dead Sea Rift Valley, in a highly faulted area. The main copper minerals are chrysocolla, malachite, pseudomalachite, bisbeite, and plancheite. These are found in cracks, veinlets, lenses, joints, bedding planes, and disseminated throughout the sediments.

c. Speckled Sandstone Formation.

(1). Location. Salt Range, Pakistan.

(2). Description. These early Permian sandstone, shale, and carbonate rocks appear to have been exposed to migrating saline ground waters, such as alkaline carbonated waters or chloride brines derived from a subsurface solution of evaporites.

d. Dzhezkazgan Carboniferous.

(1). Location. Central Kazakhstan, U. S. S. R.

(2). Description. These strata-bound copper-bearing sandstones of the Carboniferous differ significantly from red-bed deposits of other areas. Copper sulfides, with some lead, occur in gray sandstone horizons interbedded with barren brown sandstones of shallow marine lagoonal and deltaic facies. Roughly 30 mineralized zones occur within about 700 m of enclosing rock. The primary ore is sandstone impregnated with bornite, chalcocite, chalcopyrite, and galena. The first stage of mineralization appears to be syngenetic or diagenetic with the enclosing strata. This is cut by a second, epigenetic stage of Cu-Pb-Zn mineralization that forms intersecting veins, stockwork zones, and hydrothermal wallrock alteration. The ore-forming solutions may have been derived from basaltic magma at depth below the seafloor.

e. Kounrad.

(1). Location. Saline Lake Balkhash in central Kazakhstan.

(2). Description. This large porphyry deposit may eventually prove to be one of the largest Cu mining regions of the world.

f. Kalmakyr.

(1). Location. Near Tashkent, Uzbeckstan.

(2). Description. This large porphyry Cu deposit is an important source of Soviet Cu.

The Paleozoic Copper Deposits of Australasia.

The Paleozoic copper deposits of Australasia are identifiable, according to the seismotectonomagmatic-belt concept of the Tectonospheric Earth Model, with the surficial manifestations of the Earth's internal behavior that occurred along segments of the seismotectonomagmatic belts that were "active" in that area during the Paleozoic, as described in the first few pages of this chapter. To see how these predictions correlate with the facts, it is well to consider some of the Paleozoic copper deposits of that area.

It is not too surprising that some of these lie roughly parallel to some of the Mesozoic copper belts described in Chapter 7. Nor is it unexpected, according to the model, that others of these are rotatio-translationally oriented with respect to some of the Mesozoic copper deposits.

For the sake of brevity, those areas that contain Paleozoic copper deposits, as well as later deposits, are included only in the appropriate portions of Chapter 6 (Cenozoic) and/or Chapter 7 (Mesozoic).

a. Mount Morgan.

(1). Location. Queensland.

(2). Description. The 3% Cu produced from this Au-Cu deposit is mainly chalcopyrite. The irregular deposit is apparently a replacement in Carboniferous rocks, surrounded on both sides by intrusive granite.

b. Mount Lyell.

(1). Location. Tasmania.

(2). Description. The copper of this area is produced from large pyritic lenses on the west coast of Tasmania. The ore bodies replace sericite schists of apparently Paleozoic age. The main ore bodies are lenticular in shape and consist mainly of fine-grained pyrite, with 5% Cu in the form of chalcopyrite and some bornite and enargite.

Chapter 9

PROTEROZOIC SEISMOTECTONOMAGMATIC BELTS AND THE ASSOCIATED COPPER DEPOSITS

Paleomagnetic observations suggest that horizontal movements of the crust have been occurring during most of the 4.6 billion years (b. y.) that the Earth is believed to have been in existence. However, there is evidence that, during the Proterozoic, there were significant changes in the nature of the Earth's crustal movements. Whereas during the Archean the Earth's crustal masses were moving as non-rigid entities, during the Proterozoic they began to stabilize (See, e. g., Tatsch, 1972a: chapter 7; Sutton et al., 1975; Irving and McGlynn, 1975; Piper, 1975; Shackleton, 1975; Baer, 1975; Bowes, 1975; Greenwood et al., 1975; Escher et al., 1975; Kroner, 1975; Van Schmus, 1975; Watson, 1975; Zwart, 1975).

The Earth's copper deposits are identifiable, according to the seismotectonomagmatic-belt concept of the Tectonospheric Earth Model, with the surficial manifestations of the Earth's internal behavior associated with the 12 quadrispherical arcs of the model. More specifically, according to the model, the origin, evolution, and present characteristics of today's Proterozoic copper deposits are derivable primarily from the Earth's activity within the Proterozoic seismotectonomagmatic belts during that period, as modified during the Phanerozoic. The Proterozoic seismotectonomagmatic belts are now "fossil"; but they were "current", or active, during that period.

Closer scrutiny of the Tectonospheric Earth Model reveals that the Proterozoic seismotectonomagmatic belts were similar to the earlier-described belts of the Phanerozoic, in that they formed twelve 10,000-km quadrispherical arcs spanning the 6 "basic points" of the subtectonospheric fracture system of the Tectonospheric Earth Model, i. e., the 6 points that now lie roughly 1000 km beneath the Aleutians, Galapagos, Gibraltar, Bengal, Kermadecs, and Bouvet (Tatsch, 1972a: page 144). During the time that has elapsed since the Proterozoic, the crustal plates that contain today's Proterozoic copper deposits have "moved", repeatedly, according to the Tectonospheric Earth Model, with respect to these 6 "fixed" points of the model. Consequently, they have "moved" also with respect to the 12 "fixed" quadrispherical arcs that span the 6 points.

It is well to note here that the "fixed" basic points of the model are constrained to move with the "motions" of the austral and boreal poles of the model (Tatsch, 1972a: chapters 6 and 14). Even in primordial times, the Tectonospheric Earth Model had 4 boreal-to-equatorial quadrispheric arcs and 4 austral-to-equatorial quadrispheric arcs. But the surficial manifestations of the boreal and austral poles have not always been associated with the Aleutians and Bouvet islands areas as they are today (Tatsch, 1972a: chapter 17).

In view of the above, it is difficult to define, in exact terms, the post-Proterozoic "movements" of the crustal plates with respect to the 6 "fixed" points of the model. However, with these reservations in mind, it is meaningful to say that the post-Proterozoic movements of the crustal plates (with respect to any selected datum or reference) have consisted essentially of two types of motion: (1) translation and (2) rotation. The translations have been (a) in a "horizontal" (i. e., spherical-shell) domain and (b) in a "vertical" (i. e., spherical-radius) domain. The rotations, similarly, have been (c) in a "horizontal" (i. e., spherical-shell) domain and (d) in a "vertical" (i. e., central-plane) domain. These 4 types of crustal motion, previously described elsewhere (Tatsch, 1972a), may be reviewed briefly here.

Of particular interest to analyses of copper deposits are the crustal-plate movements that combine components from two or more of the 4 types of the listed rotatio-translational motion (i. e., a, b, c, and d, above). The geometrical, mechanical, thermal, and chemical aspects of these crustal-plate movements are provided by the three (and four) dimensional omniductive plate motions defined by the Tectonospheric Earth Model (Chapter 4).

Although a discussion of the details of the higher-order applications of these motions is beyond the scope of this book, some of these details have been discussed elsewhere (Tatsch, 1972a: chapters 3, 6, 7, 15, 16, 17, and 18; 1973a). Also, a comprehensive appreciation for the causes and effects of the rotating crustal plates is provided elsewhere (Tatsch, 1972a: figure 17-10-2-3, on page 787, and the accompanying text there). Suffice it here to summarize the aspects of the crustal-plate motions, and their causes, that have a bearing on the origin, evolution, and present characteristics of today's Proterozoic copper deposits.

Briefly stated, an analysis of the geometrical, mechanical, thermal, and chemical aspects of the Earth's behavior during the post-Proterozoic period, as interpreted by the Tectonospheric Earth Model, suggests that there are certain differences between today's Proterozoic and Cenozoic copper deposits. These differences may be observed by considering (1) the origin of today's Proterozoic deposits during the roughly 2 b. y. that span the Proterozoic; (2) the geometrical, mechanical, thermal, and chemical aspects of the Earth's post-Proterozoic behavior that have modified the Proterozoic crustal plates and have affected, thereby, the evolution of today's Proterozoic copper deposits; and (3) the present characteristics of today's Proterozoic copper deposits as a function of the Earth's post-Proterozoic behavior.

The origin, evolution, and present characteristics of today's Proterozoic copper deposits are associated, as indicated above, with the seismotectonomagmatic belts that developed within the upper 1000 km of the Earth during the Proterozoic (Tatsch, 1972a: page 269 and chapter 7). Data bearing on this period of the Earth's history have been analyzed elsewhere (See, e. g., Brock, 1973; Burk, 1975; Creer, 1973; Dalziel et al., 1973; Deutsch and Murthy, 1973; Dunnett, 1975; Fahrig, 1971; Goodwin, 1973; Greenwood et al., 1975; Hills et al., 1975; Hurley, 1973; Johnson, 1973; Kisvarsanyi, 1974; Lowry, 1974; Morton et al., 1974; Palmer, 1974; Piper, 1973a; Roaldset, 1975; Rolands, 1975; Shackleton, 1973; and Sutton, 1973). The results in many of these analyses suggest that the Earth's post-Proterozoic behavior resulted from stress systems that began in the Precambrian (See, e. g., Lowry, 1974).

The Proterozoic seismotectonomagmatic belts are now "fossil", but it is well to take a brief look at how they might have looked during the early Proterozoic. Basically, there were then, according to the model, twelve 10,000-km quadrispherical arcs spanning the 6 "fixed" basic points of the subtectonospheric fracture system, i. e., the 6 points that now lie roughly 1000 km beneath the Aleutians, Galapagos, Gibraltar, Bengal, Kermadecs, and Bouvet, as described earlier. Thus, there were, during the early Proterozoic, 3 then-current seismotectonomagmatic belts, similar to today's Quaternary, or now-current, seismotectonomagmatic belts. Each of the twelve 10,000-km quadrispherical arcs of the Proterozoic was associated, similarly, with two of the 6 "fixed" basic points of the subtectonospheric fracture system of the Tectonospheric Earth Model. For example, one of the 10,000-km quadrispherical arcs that is associated today with the Proterozoic copper deposits of the Canadian Shield was associated, during the Proterozoic also, with a segment of the equatorial seismotectonomagmatic belt that is now identifiable with a great circle passing roughly through the Galapagos, Gibraltar, Bengal, and Kermadecs (Tatsch, 1972a: page 144).

On the surface _of the Earth_, the Proterozoic quadrispherical arc described above is still roughly where it was during the Proterozoic. But, with respect to the "fixed" equatorial belt (roughly 1000 km beneath the Galapagos, Gibraltar, Bengal, and Kermadecs) on the surface of the subtectonosphere, the Proterozoic arc on the surface of the Earth has undergone rotational and translational motions during the time that has elapsed since the early Proterozoic.

Today's Proterozoic copper deposits, according to the Tectonospheric Earth Model, originated in the subsurficial arcs while the substances genetic to these deposits were roughly radially above various segments of the corresponding arcs that span the 6 "fixed" basic points on the surface of the subtectonosphere. Thus, today's Proterozoic copper deposits that lie within the Canadian Shield, for example, were formed while each applicable area lay roughly radially above corresponding segments of the seismotectonomagmatic belts that were "active" during the Proterozoic.

Similar conclusions may be drawn regarding today's Proterozoic copper deposits that lie in other parts of the world. To see how well these concepts apply to the Earth's actual Proterozoic copper deposits, some representative Proterozoic copper deposits of each continent are analyzed in the following sections.

The Proterozoic Copper Deposits of North America.

The Proterozoic copper deposits of North America are identifiable, according to the seismotectonomagmatic-belt concept of the Tectonospheric Earth Model, with the surficial manifestations of the Earth's internal behavior that occurred along segments of the seismotectonomagmatic belts that were "active" in that area during the Proterozoic, as described in the first few pages of this chapter. To see how these predictions correlate with the facts, it is well to consider some of the Proterozoic copper deposits of that area.

It is not too surprising that some of these lie roughly parallel to some of the Paleozoic copper belts described in Chapter 8. Nor is it unexpected, according to the model, that others of these are rotatio-translationally oriented with respect to some of the Paleozoic copper belts.

As in other continents, the Proterozoic seismotectonomagmatic belts of North America differ somewhat from the earlier Archean belts (See, e. g., Tatsch, 1972a: chapter 7). These differences, as well as some similarities, are particularly evident in the Circum-Ungava belt of Canada (Dimroth et al., 1972). In this belt, a miogeosyncline, adjoining a well-defined craton, passes outward in a eugeosyncline accompanied by increasing intensity of metamorphism.

For the sake of brevity, those areas that contain Proterozoic copper deposits, as well as later deposits, are included only in the appropriate portions of Chapter 6 (Cenozoic), Chapter 7 (Mesozoic), and/or Chapter 8 (Paleozoic).

a. Coppermine.

(1). Location. Northwest Territories, Canada.

(2). Description. Resembling the Keweenaw deposits of Michigan, these chalcocite-bearing deposits lie in gently dipping basalt flows of Precambrian-Y age.

b. Belt Supergroup.

(1). Location. Western Montana and adjacent parts of Idaho.

(2). Description. Copper sulfides occur as disseminations, blebs, and veinlets in persistent beds of white fine-grained quartzite and siltite of the Revett Formation along a belt parallel to a broad domal feature formed before the close of the Beltian, perhaps as early as one b. y. ago. Anomalous amounts of

Cu are more widely found in the green strata of the Belt basin. The apparent relation between mineralization and a late Proterozoic broad domal structure suggests a diagenetic or early epigenetic origin for the Cu (See, e. g., Harrison, 1972). Later recrystallization and metamorphism to the biotite zone reduced the permeability, thereby precluding the possibility of late epigenetic mineralization.

c. Jerome-Bradshaw Mountains.

(1). Location. Arizona.

(2). Description. These pyritic replacement deposits are contained in Precambrian schists of sedimentary and igneous origin. Their development followed the intrusion of the Precambrian Bradshaw granite. The deposit forms a pipe 100 m in diameter and at least 1 km deep. The fine-grained ore replaces rhyolite schist and contains about 5% Cu. The pyrite is followed by sphalerite, chalcopyrite, and tennantite. No barite nor distinctive high-temperature minerals are present.

d. Manitoba.

(1). Location. Canada.

(2). Description. These sulfides, often massive, comprise lenticular replacement bodies of pyrite, pyrrhotite, chalcopyrite, and sphalerite. They occur in shear zones of gneiss or in Keewatin volcanics. The Sherritt-Gordon and Flin-Flon mines are part of this complex.

e. Duluth Gabbro Complex.

(1). Location. Northern Minnesota.

(2). Description. These disseminated copper and nickel sulfide deposits form discontinuous lenses and layers in the basal part of the Precambrian gabbro complex. Although not now economically minable, they represent a significant future copper resource.

f. Keweenaw Peninsula.

(1). Location. Michigan.

(2). Description. These deposits of native copper occur in amygdaloidal flow tops and conglomerate beds in the Portage Lake Volcanics. These strata-bound and partially conglomerate-bed-controlled ores are usually considered to be epigenetic and to be related to some strata-bound ores in sedimentary rocks, such as the mineralized Nonesuch Shales above the Portage Lake Volcanics. However, the widespread minor Cu occurrence in basaltic rocks all over the world suggests that the Cu was originally syngenetic but has been redistributed and concentrated by later processes. Most of the copper is native metal that was introduced into vesicles, fractures, and interstitial pore spaces after deformation of the rocks. The deposits appear to be related to a pattern of low-grade regional metamorphic zoning of silicate minerals. The source of the mineralizing solutions for these deposits may be essentially the same as that for the copper sulfide deposits in the overlying Nonesuch Shale, i. e., the 5 to 10 km of flood basalts within the Lake Superior basin. White (1968) feels that the copper-bearing solutions may represent water driven from the voids and interstices of low tops as the basalt flows were crushed and metamorphosed in the axial region of the Lake Superior syncline, downdip from the present deposits.

g. White Pine.

(1). Location. Ontonogan County, Michigan.

(2). Description. Roughly 5% of the United States copper production comes from this Proterozoic Nonesuch Shale found primarily in basal parting shale beds. Thicknesses range from 1 to

8 m with anomalous copper over a distance of at least 250 km in the Upper Peninsula of Michigan and northern Wisconsin. The primary Cu mineral is chalcocite, the richest ores occurring in the darkest carbonaceous siltstones, shales, and sandstones. A crosscutting relation between ore and bedding on a regional scale and the absence of a characteristic copper-bearing lithologic facies distinguish this deposit from the African copper belt. The most likely source is epigenetic non-magmatic, but some evidence suggests a syngenetic source (See, e. g., Cox et al., 1973).

h. Calumet and Heckla.

(1). Location. Michigan.

(2). Description. These copper deposits comprise two broad groups: (1) lodes and (2) fissures. The lodes are conglomerates of felsite inerstratified with the lavas and amygdaloidal vesicles and breccias from the tops of lava flows. The fissures are veins along the fractures that parallel or cross the beds. Most of the good ores are confined to the Keweenawan series composed primarily of lava flows. The Nonesuch lode is a copper-bearing sediment that occurs well up in the sedimentary portion of the series where the mineralization is mainly in sandstone and, to a lesser extent, in shale.

i. Sudbury.

(1). Location. Sudbury, Ontario, Canada.

(2). Description. Copper is produced as a by-product of nickel-bearing mafic rocks. These stratiform and crosscutting ores within complex geology are polygenic with respect to the enclosing rocks. However, they were probably derived initially from crystallization and the separation of a dense sulfide-rich liquid. Similar Cu segregations in mafic intrusive bodies are found in Maine, New Brunswick, Quebec, Manitoba, Saskatchewan,

British Columbia, Alaska, Norway, Sweden, Finland, the Kola Peninsula, Siberia, and other locations aligned along these Proterozoic seismotectonomagmatic belts.

j. Chibougamau.

(1). Location. Quebec.

(2). Description. These sulfide lenses, in shears crosscutting anorthosite bodies, do not fit into any of the classical types of copper deposits. Control appears to have been from the intersection of two major shear zones.

k. Catoctin Formation.

(1). Location. Pennsylvania, Virginia, and Maryland.

(2). Description. These Keweenaw-resembling deposits of native copper, cuprite, azurite, and malachite lie in chloritized and epidotized mafic flows of Precambrian-Z age. Partly amygdaloidal and partly schistose, the containing basalts are irregularly fractured.

The Proterozoic Copper Deposits of South America.

The Proterozoic copper deposits of South America are identifiable, according to the seismotectonomagmatic-belt concept of the Tectonospheric Earth Model, with the surficial manifestations of the Earth's internal behavior that occurred along segments of the seismotectonomagmatic belts that were "active" in that area during the Proterozoic, as described in the first few pages of this chapter. To see how these predictions correlate with the facts, it is well to consider some of the Proterozoic copper deposits of that area.

It is not too surprising that some of these lie roughly parallel to some of the Paleozoic copper belts described in Chapter 8.

Nor is it unexpected, according to the model, that others of these are rotatio-translationally oriented with respect to some of the Paleozoic copper belts.

For the sake of brevity, those areas that contain Proterozoic copper deposits, as well as later deposits, are included only in the appropriate portions of Chapter 6 (Cenozoic), Chapter 7 (Mesozoic), and/or Chapter 8 (Paleozoic).

a. *Brazilian Shield.*

(1). *Location.* Brazil.

(2). *Description.* Unknown quantities of Cu are associated with the Brazilian Shield.

The Proterozoic Copper Deposits of Africa.

The Proterozoic copper deposits of Africa are identifiable, according to the seismotectonomagmatic-belt concept of the Tectonospheric Earth Model, with the surficial manifestations of the Earth's internal behavior that occurred along segments of the seismotectonomagmatic belts that were "active" in that area during the Proterozoic, as described in the first few pages of this chapter. To see how these predictions correlate with the facts, it is well to consider some of the Proterozoic copper deposits of that area.

It is not too surprising that some of these lie roughly parallel to some of the Paleozoic copper belts described in Chapter 8. Nor is it unexpected, according to the model, that others of these are rotatio-translationally oriented with respect to some of the Paleozoic copper belts.

For the sake of brevity, those areas that contain Proterozoic copper deposits, as well as later deposits, are included only in the appropriate portions of Chapter 6 (Cenozoic), Chapter 7 (Mesozoic), and/or Chapter 8 (Paleozoic).

a. Zambia-Katanga.

(1). Location. Zambia and the Katanga province of Zaire.

(2). Description. Roughly 20% of the world's copper production comes from Proterozoic sedimentary rocks that extend at least 500 km across parts of Zambia and the Katanga province of Zaire. The uncertainty of their origin causes these deposits to be labeled with the non-descript term "stratabound". The ore occurs in dark shales of the Lower Roan Formation, most of which comprises arkosic conglomerates and eolian sandstones overlying a granitic basement. The ore horizon was probably deposited in a shallow, restricted basin surrounded by algal reefs. A syngenetic origin is suggested by the zonal arrangement of the ore minerals parallel to the shoreline as it migrated back and forth with transgressions and regressions of the sea. From barren sediments along the shoreline, the sequence proceeds through disseminated chalcocite, bornite, and chalcopyrite, to pyrite in the same general direction of current flow. The mineralized zone cuts stratigraphically upward through conglomerates, sandstones, argillites, and dolomites. Actually, the African copper belt may have resulted from a combination of several processes: (1) on the assumption that the deposits are syngenetic, they may have formed during an interval of time when oxygen first became available in sufficient quantity to mobilize the Cu, i. e., about 1.3 b. y. ago; (2) a marine transgression may have mobilized the Cu in the underlying non-marine crystalline rocks suggestive of the Kupferschiefer (Wedepohl, 1971; Putzer, 1972). The African copper belt is estimated to contain at least one billion tons of ore (See, e. g., Cox et al., 1973).

b. Messina.

(1). Location. Transvaal.

(2). Description. These hypogene chalcocite, bornite, and chalcopyrite ores occur in granite gneiss altered to zoisite quartz.

c. Ansirabe.

(1). Location. Madagascar.

(2). Description. Significant Precambrian mineral deposits are found at the intersection of the Mahanoro-Tsiribihina fracture zone with the cross-cutting axis of magmatic reactivation (See, e. g., Kutina, 1975). The copper deposits extend along the axis of magmatic reactivation and represent the most striking structural control of ore deposition in Madagascar. Both west and east of the axis of reactivation, other deposits are found along equally spaced lines parallel to the axis of magmatic reactivation. The equal spacing of these lines suggests a set of north-northwest-trending, deep-seated fractures or zones of weakness, such as would be expected within a seismotectonomagmatic belt.

The Proterozoic Copper Deposits of Europe.

The Proterozoic copper deposits of Europe are identifiable, according to the seismotectonomagmatic-belt concept of the Tectonospheric Earth Model, with the surficial manifestations of the Earth's internal behavior that occurred along segments of the seismotectonomagmatic belts that were "active" in that area during the Proterozic, as described in the first few pages of this chapter. To see how these predictions correlate with the facts, it is well to consider some of the Proterozoic copper deposits of that area.

It is not too surprising that some of these lie roughly parallel to some of the Paleozoic copper belts described in Chapter 8. Nor is it unexpected, according to the model, that others of these are rotatio-translationally oriented with respect to some of the Paleozoic copper belts.

For the sake of brevity, those areas that contain Proterozoic copper deposits, as well as later deposits, are included only in the appropriate portions of Chapter 6 (Cenozoic), Chapter 7 (Mesozoic), and/or Chapter 8 (Paleozoic).

a. Aitik.

(1). Location. Gallivare, Sweden.

(2). Description. These low-grade Cu ores appear to have been sedimentary in origin, with subsequent modification by metamorphism. Deposited within Precambrian gneisses, schists, and quartzites the Cu is primarily stringers and disseminations of chalcopyrite mixed with pyrite, magnetite, and pyrrhotite. Some hydrothermal alteration accompanied the mobilization of the sulfides.

b. Boliden-Skellefte.

(1). Location. Sweden.

(2). Description. This large lens of Au-Ag-Cu-As-S deposits has produced important quantities of copper. Some copper came also from the pyritic masses at Falun. The steep lenses attain lengths of 600 m and are enclosed in an Archean supracrustal formation of acidic flows and tuffs overlain by phyllite and quartzite. Modern analyses suggest that these deposits were formed in association with volcanism in a shallow marine environment during the formation of a geosyncline and that they subsequently were subjected to at least two periods of deformation (See, e. g., Rickard and Zweifel, 1975).

c. Kola Peninsula.

(1). Location. European Russia.

(2). Description. These copper pentlandite sulfides are similar to those in other parts of the Soviet Arctic,

i. e., Norylsk, Taimyr Peninsula, and Koryak Peninsula.

d. Autokumpo.

(1). Location. Finland.

(2). Description. Lying beneath a swamp, this ore body is in brecciated quartzite. The breccia fragments are joined by chalcopyrite and pyrrhotite and are invaded by pyrite and chalcopyrite.

The Proterozoic Copper Deposits of Asia.

The Proterozoic copper deposits of Asia are identifiable, according to the seismotectonomagmatic-belt concept of the Tectonospheric Earth Model, with the surficial manifestations of the Earth's internal behavior that occurred along segments of the seismotectonomagmatic belts that were "active" in that area during the Proterozoic, as described in the first few pages of this chapter. To see how these predictions correlate with the facts, it is well to consider some of the Proterozoic copper deposits of that area.

It is not too surprising that some of these lie roughly parallel to some of the Paleozoic copper belts described in Chapter 8. Nor is it unexpected, according to the model, that others of these are rotatio-translationally oriented with respect to some of the Paleozoic copper belts.

For the sake of brevity, those areas that contain Proterozoic copper deposits, as well as later deposits, are included only in the appropriate portions of Chapter 6 (Cenozoic), Chapter 7 (Mesozoic), and/or Chapter 8 (Paleozoic).

a. Udokan.

(1). Location. Eastern Siberia, U. S. S. R.

(2). Description. This multi-million-ton copper deposit comprises undeveloped sedimentary ores that could easily make it the world's largest sedimentary copper deposit (Shank, 1972). The ores, mainly sedimentary and lithologic, contain the greatest Cu concentrations in the 1.8-to-2.5-km-thick Sakukan Formation within the 11-km Udokan Group of fine sandstone and siltstone and some thin carbonate lenses. Deposition was in shallow water or in a littoral-deltaic environment. The Cu was deposited with the sedimentary material or diagenetically, probably from an older Archean shield area to the northeast. Transport was probably as fine clastic particles. The Udokan Series comprises a large synclinal folded basin, irregularly oval in plan, that is slightly metamorphosed. The main ores are disseminated chalcocite and bornite plus some apparently unrelated chalcopyrite ores. Malachite, brochantite, and other secondary minerals occur in a restricted oxidized zone.

b. Taimyr Peninsula.

(1). Location. Soviet Arctic.

(2). Description. These ores contain commercial concentrations of both Cu and Ni, and they are treated for the recovery of the platinum group metals.

c. Norylsk.

(1). Location. Siberia.

(2). Description. These ores are similar to the deposits of the Taimyr Peninsula, but they are farther to the north.

The Proterozoic Copper Deposits of Australasia.

The Proterozoic copper deposits of Australasia are identifiable, according to the seismotectonomagmatic-belt concept of the Tectonospheric Earth Model, with the surficial manifestations of the Earth's internal behavior that occurred along segments of the seismotectonomagmatic belts that were "active" in that area during the Proterozoic, as described in the first few pages of this chapter. To see how these predictions correlate with the facts, it is well to consider some of the Proterozoic copper deposits of that area.

It is not too surprising that some of these lie roughly parallel to some of the Paleozoic copper belts described in Chapter 8. Nor is it unexpected, according to the model, that others of these are rotatio-translationally oriented with respect to some of the Paleozoic copper belts.

For the sake of brevity, those areas that contain Proterozoic copper deposits, as well as later deposits, are included only in the appropriate portions of Chapter 6 (Cenozoic), Chapter 7 (Mesozoic), and/or Chapter 8 (Paleozoic).

a. Mount Isa.

(1). Location. Queensland, Australia.

(2). Description. These Cu ores appear to have been sedimentary in origin, with later modification by metamorphism. Associated with Pb-Zn-Ag ores that form masses conformable with the bedding of the Precambrian Mount Isa Shale, the Cu ores occur in shears, particularly where silica and dolomite have been introduced. Thus, although the Pb-Zn-Ag ores appear to be syngenetic, the origin of the associated Cu sulfides is less certain.

b. Kambalda District.

(1). Location. Western Australia.

(2). Description. These Ni ores in Precambrian mafic andultramafic rocks average roughly 0.4% copper.

c. Juno.

(1). Location. About 960 km south of Darwin,

Northern Territory, Australia.

(2). Description. Native gold, chalcopyrite, and various bismuth sulfosalts are concentrated in two Proterozoic ellipsoidal-to-pipelike bodies of magnetite that comprise the Juno complex.

Chapter 10

ARCHEAN SEISMOTECTONOMAGMATIC BELTS AND THE ASSOCIATED COPPER DEPOSITS

For the purpose of this analysis, it is assumed, that the Earth's surface 4.6 billion years (b. y.) ago was devoid of any surficial features other than those that would have been genetic to an accretionary process of planetary evolution from the solar nebula at that time (Tatsch, 1972a: chapter 1). The evolutionary behavior of the mineral deposits, during the early Archean, would have been motivated by the Earth's internal behavior, as influenced by those extraterrestrial factors that might have been operating upon the Earth at that time.

According to the Tectonospheric Earth Model, the fragments of the 5 "terrestrial" octants of Earth Prime became accreted onto the 5400-km primordial Earth during the early part of the earliest "Archean", i. e., during the period 4.6 b. y. to 3.6 b. y. ago. Present evidence does not define clearly the exact time at which this accretion occurred within this one-billion-year span. Nor is it necessary to this discussion that this be known exactly. Suffice it here to say merely that the evolution of the copper deposits during the period 4.6 b. y. to 3.6 b. y. ago depended primarily upon two basic aspects of the Earth's behavior during that period: (1) the mineral "mix" within the accreting fragments of Earth Prime was not as homogeneous as was that of the surface of the 5400-km primordial Earth; and (2) the primordial Earth, as it accreted to its present size, behaved in accordance with the geometrical, mechanical, thermal, and chemical aspects described in Chapters 2 and 3.

It is of interest to note, in this connection, that many of these accreting fragments bore a marked resemblance to the rocks that now comprise the Moon, the asteroids, and the meteorites (Tatsch, 1972a, 1973a, 1974a).

The above-described, early-Archean behavior of the Earth produced the following constraints upon the evolutionary behavior of the copper deposits during the period 4.6 b. y. to 3.6 b. y. ago: (1) the surface of the 5400-km primordial Earth was sialic material cut by three mutually-orthogonal belts of extruding mafic and ultramafic material; and (2) the accreting material (from the fragments of the 5 "terrestrial" octants of Earth Prime) comprised a complete spectrum of material, ranging from (a) fragments of iron, nickel-iron, and other metals that might have been contained within the molten and plastic parts of Earth Prime prior to its fragmentation, to (b) sialic material (from the surface of Earth Prime) which was similar to that on the surface of the 5400-km primordial Earth (Tatsch, 1973a).

These constraints produced the following evolutionary behavior within the Earth's surface during the period 4.6 b. y. to 3.6 b. y. ago: (1) the initial sialic minerals on the surface of the primordial 5400-km Earth and on the erstwhile surface of Earth Prime were thoroughly metamorphosed by the effects of the accretionary activity and of the geometrical, mechanical, thermal, and chemical behavior of the primordial Earth as it grew from about 5400 km to its present size; and (2) as the size of the primordial Earth increased (at an average rate of about one millimeter per year), the geometrical, mechanical, thermal, and chemical effects of the Earth's internal behavior were evidenced primarily within the primordial orogenic belts of the Earth during the period 4.6 b. y. to 3.6 b. y. ago, i. e., through the seismotectonomagmatic belts of the primordial Earth.

This evolutionary behavior within the Earth's surface (as it grew at an average rate of a few micrometers per day) produced the following effects upon the distribution patterns of the mineral deposits during that period: (1) the igneous, metamorphic, and sedimentary processes associated with mineral deposits tended to form three mutually-orthogonal belts upon the surface of the Earth; and (2), as the Earth grew, at the rate of a few micrometers per day, it tended to continually equilibrate its condition to one of minimum energy by modifying the upper few micrometers of its surface each day, through the geometrical, mechanical, thermal, and chemical aspects of its driving mechanism (Chapters 2 and 3), thereby redistributing, metamorphosing, and otherwise affecting the mineral deposits to the extent that would be expected as a result of the geometrical, mechanical, thermal, and chemical effects of the basic driving mechanism under these conditions (Tatsch, 1972a, 1973a).

The most probable distribution pattern of the mineral deposits during the middle Archean (i. e., about 3.6 b. y. ago) may be summarized: (1) the youngest mineral belts (i. e., those 3.6 b. y. old on the present time scale) were aligned along various segments of three mutually-orthogonal belts; and (2) the older mineral belts consisted of "fossil" belts that had been overprinted, metamorphosed, and otherwise changed to the extent that would be expected as a result of the geometrical, mechanical, thermal, and chemical effects of the Earth's internal behavior during the earliest of the early Archean, i. e., during the period 4.6 b. y. to 3.6 b. y. ago (Tatsch, 1972a, 1973a).

<u>The Earth's Archean Copper Deposits</u>.

From the above, it may be seen that the Earth's Archean copper deposits are identifiable, according to the seismotectonomagmatic-belt concept of the Tectonospheric Earth Model, with the surficial manifestations of the Earth's internal activity associated with the 12 quadrispherical arcs of the model, quite similarly to the manner described in earlier chapters for the Cenozoic, Mesozoic, Paleozoic, and Proterozoic. More specifically, according to the model, the origin, evolution, and present characteristics of today's Archean copper deposits are derivable primarily from the Earth's activity within the Archean seismotectonomagmatic belts during that period, as modified during the Proterozoic and the Phanerozoic. As indicated earlier, the Archean seismotectonomagmatic belts are now "fossil"; but they were "current", or active, during that period.

Closer scrutiny of the Tectonospheric Earth Model reveals that the Archean seismotectonomagmatic belts were similar to the earlier-described belts of the Proterozoic, in that they formed twelve 10,000-km quadrispherical arcs spanning the 6 "basic points" of the subtectonospheric fracture system of the Tectonospheric Earth Model, i. e., the 6 points that now lie roughly 1000 km beneath the Aleutians, Galapagos, Gibraltar, Bengal, Kermadecs, and Bouvet (Tatsch, 1972a: page 144).

One salient difference between the earliest Archean Earth and the later Earth was the condition of the Earth's surface (See, e. g., Brock, 1973; Coward, 1975; Creer, 1973; Goodwin, 1973, 1975; Hawkesworth et al., 1975a, 1975b; Hurley, 1973; Morioka and Kigoshi, 1975; Schau, 1975; Sutton, 1973; Tatsch, 1972a; Veizer, 1975). Briefly stated, the Earth's surface during the earliest Archean was thinner and more pliable, perhaps, than it is today. The concept of "crustal plates", as known today, was not applicable during the earliest Archean.

During the time that has elapsed since the Archean, the crustal plates that contain today's Archean copper deposits have "moved", repeatedly, according to the Tectonospheric Earth Model, with respect to the 6 "fixed" points of the model. Consequently, the crustal plates have "moved" also with respect to the 12 "fixed" quadrispherical arcs that span the 6 points.

It is well to note here that the "fixed" basic points of the model are constrained to move with the "motions" of the austral and boreal poles of the model (Tatsch, 1972a: chapters 6 and 14). Even in the earliest Archean times, the Tectonospheric Earth Model had 4 boreal-to-equatorial quadrispheric arcs and 4 austral-to-equatorial quadrispheric arcs. But the surficial manifestations of the boreal and austral poles have not always been associated with the Aleutians and Bouvet islands, as they are today (Tatsch, 1972a: chapter 17).

In view of the above, it is difficult to define, in exact terms, the post-Archean "movements" of the crustal plates with respect to the 6 "fixed" points of the model. However, with these reservations in mind, it is meaningful to say that the post-Archean movements of the crustal plates (with respect to any selected datum or reference) have consisted essentially of two types of motion: (1) translation and (2) rotation. The translations have been (a) in a "horizontal" (i. e., spherical-shell) domain and (b) in a "vertical" (i. e., spherical-radius) domain. The rotations, similarly, have been (c) in a "horizontal" (i. e., spherical-shell) domain and (d) in a "vertical" (i. e., central-plane) domain. These 4 types of crustal motion, previously described elsewhere (Tatsch, 1972a), may be reviewed briefly here.

Of particular interest to analyses of copper deposits are the crustal-plate motions that combine components from two or more of the 4 types of the listed rotatio-translational motion (i. e., a, b, c, and d, above). The geometrical, mechanical, thermal, and chemical aspects of these crustal-plate motions are provided by the three (and four) dimensional omniductive plate motions defined by the Tectonospheric Earth Model (Chapter 4).

Although a discussion of the details of the higher-order applications of these motions is beyond the scope of this book, some of these details have been discussed elsewhere (Tatsch, 1972a: chapters 3, 6, 7, 15, 16, 17, and 18; 1973a). Also, a comprehensive appreciation for the cause and effects of the rotating crustal plates is provided elsewhere (Tatsch, 1972a: figure 17-10-2-3, on page 787, and the accompanying text there). Suffice it here to summarize the aspects of the crustal-plate motions and their causes that have a bearing on the origin, evolution, and present characteristics of today's Archean copper deposits.

B Briefly stated, an analysis of the geometrical, mechanical, thermal, and chemical aspects of the Earth's behavior during the post-Archean period, as interpreted by the Tectonospheric Earth Model, suggests that there are certain differences between today's Archean and Cenozoic copper deposits. These differences may be observed by considering (1) the origin of today's Archean deposits during the roughly 2 b. y. that span the Archean; (2) the geometrical, mechanical, thermal, and chemical aspects of the Earth's post-Archean behavior that have modified the Archean crustal plates and have affected, thereby, the evolution of today's Archean copper deposits; and (3) the present characteristics of today's Archean copper deposits as a function of the Earth's post-Archean behavior.

The origin, evolution, and present characteristics of today's Archean copper deposits are associated, as indicated above, with the seismotectonomagmatic belts that developed within the upper 1000 km of the present Earth during the Archean (Tatsch, 1972a: page 269 and chapter 7). Data bearing on this period of the Earth's history have been analyzed elsewhere (See, e. g., Brock, 1973; Creer, 1973; Goodwin, 1973; Hurley, 1973; Johnson, 1973; Lowry, 1974; Palmer, 1974; and Sutton, 1973). The results in many of these analyses suggest that the Earth's post-Archean behavior resulted from stress systems that began early in the history of the Earth (See, e. g., Lowry, 1974).

Today's Archean copper deposits, according to the Tectonospheric Earth Model, originated in the subsurficial quadrispheric arcs of the model while the substances genetic to these deposits were roughly radially above various segments of the corresponding arcs that span the 6 "fixed" basic points on the surface of the subtectonosphere. Thus, today's Archean copper deposits that lie within the Canadian Shield, for example, were formed while each applicable area lay roughly radially above corresponding segments of the seismotectonomagmatic belts that were "active" during the Archean. Some of the constraints on analyses of this type have been discussed by Hayatsu and Palmer (1975), as well as by the researchers cited earlier in this section.

Similar conclusions may be drawn regarding today's Archean copper deposits that lie in other parts of the world. However, the paucity of observational data regarding the Archean precludes a complete global analysis, continent by continent, of the Earth's Archean copper deposits. Suffice it to say merely that what data there are appear to be consistent with the predictions of the model for the Archean period of the Earth's development.

Chapter 11

THE PRESENT DISTRIBUTION OF COPPER DEPOSITS AS A FUNCTION OF THE EARTH'S BEHAVIOR DURING THE PAST 4.6 BILLION YEARS

Before discussing the use of the Tectonospheric Earth Model as a supplementary tool in the exploration for copper deposits, it is well to review briefly the origin, evolution, and present characteristics of these deposits on a global basis. This is best done by considering the deposits of various ages within each of the continents. No purpose is served in such an analysis by fabricating various "pre-drift" reconstructions for various epochs of the Mesozoic, Paleozoic, Proterozoic, and Archean. Suffice it here to say merely that, during the 4.6 b. y. that the Earth has been in existence, any copper mineral now at the surface of the Earth could have circumnavigated the globe at least once. Thus, any copper mineral could have gone completely around the Earth and returned to its original place. Or it could have gone half way around and then returned to its original location. Or it could have journeyed a highly circuitous route to arrive at its present location. The concept of continental drift permits many possibilities during the past 4.6 b. y. The concept of the Tectonospheric Earth Model places certain geometrical, mechanical, thermal, and chemical constraints on those possibilities. These need not be considered here but they have been discussed elsewhere (See, e. g., Tatsch, 1972a). By summarizing the present positions of the copper deposits within each continent, the reader is permitted to make his own "pre-drift" reconstructions for each epoch of the Archean, Proterozoic, and Phanerozoic. Or the reader may, if he wishes, leave all such constructions to others. In applying the Tectonospheric Earth

Model as a supplementary tool in the exploration for copper deposits, it is largely immaterial whether or not such reconstructions are made, or how they are made, so long as the observational evidence is not distorted by any reconstructions that are undertaken.

Subject to the above comments, the patterns of copper-deposit emplacements that are predictable from the Tectonospheric Earth Model may be summarized within each continent:

a. North America. According to the model, the emplacement of copper deposits is closely associated with the positions of seismotectonomagmatic belts. We see, for example, that the eastern part of North America contains few copper deposits younger than the Mesozoic, because no active seismotectonomagmatic belts have been associated with that part of North America since that time. On the other hand, the western part of North America contains a wealth of copper deposits associated with the present (Cenozoic) position of a segment of one of the seismotectonomagmatic belts of the model. Similarly, the Archean, Proterozoic, Paleozoic, and Mesozoic copper deposits described in Chapter 7, 8, 9, and 10, are associated with the positions of segments of the applicable seismotectonomagmatic belts during those times. The copper within the Keweenawan igneous rocks is associated, according to this concept, with the Proterozoic position of a segment of the applicable seismotectonomagmatic belt. Similarly, the early Paleozoic copper deposits running from Nova Scotia to Alabama are associated with the early Paleozoic position of a segment of the applicable seismotectonomagmatic belt. The feeble recurrence of copper depositions in the Triassic traps of New Jersey, Connecticut, and the Bay of Fundy are associated with the early Mesozoic position of a segment of the applicable seismotectonomagmatic belt. Copper deposition in other parts of the North America may be analyzed in a similar manner. For example, there are very few Proterozoic copper deposits along the western part of North America because no active segments of a seismotectonomagmatic belt were associated with that area at that time (Chapter 4).

b. South America. Much of what has been said in the previous section about North America applies also to South America.

c. Africa. Most of the copper deposits of southern Africa are associated with Mesozoic and older positions of segments of the applicable seismotectonomagmatic belts. In the north, i. e., in Morocco, Algeria, and Tunisia, the few copper deposits there are associated with the present (Cenozoic) position of the Gibraltar-to-Bengal segment of the equatorial seismotectonomagmatic belt.

d. Europe. Most of the Precambrian copper deposits of Europe are associated with the Precambrian positions of segments of the applicable seismotectonomagmatic belts, i. e., in the Fennoscandia shield of Sweden, Norway, and Finland. The copper deposits within the metamorphosed Cambro-Silurian of northern Norway, Great Britain, France, Germany, Czechoslovakia, and Spain are associated with the early Paleozoic position of segments of the applicable seismotectonomagmatic belts. Later in the Paleozoic and in the early Mesozoic, the copper deposits of England, France, Spain, Germany, and the Ural Mountains are associated with segments of the seismotectonomagmatic belts that produced the Hercynian and related orogenies. During the Cenozoic, the copper deposits of southern Europe are associated with the present (Cenozoic) position of the Gibraltar-to-Bengal segment of the equatorial seismotectonomagmatic belt.

e. Asia. Within the Indian and Siberian shields, the Precambrian copper deposits are associated with the Precambrian positions of the segments of the applicable seismotectonomagmatic belts. These include the Mysore belt and belts on either side of Lake Baikal. Paleozoic copper deposits of the Urals, the Altai, and the Trans-Baikalian mountains are associated with the Paleozoic positions of the applicable seismotectonomagmatic belts. The same may be said about the copper deposits within the Permian sandstones of northern Turkestan and the older copper deposits of Malaysia, Burma, China, and Siam. The Cenozoic copper deposits along the eastern coast of Asia are associated with the present (Cenozoic) position of the Bengal-to-Aleutians segment of the meridional seismotectonomagmatic belt that passes also through the Galapagos and Bouvet.

f. Australasia. Because no segments of any of the three seismotectonomagmatic belts have been assoicated with the Australian mainland since the Mesozoic, very few Cenozoic copper deposits are found in Australia. The South Australian copper deposits of Wallaroo, Moonta, and Burra Burra are associated with the Precambrian position of a segment of the applicable seismotectonomagmatic belt. The Paleozoic copper deposits in Victoria, in the adjacent parts of New South Wales, and in Tasmania are associated with the early Paleozoic position of segments of the applicable seismotectonomagmatic belts. Later in the Paleozoic and in the early Mesozoic, the copper deposits of the New England region, in the northern part of New South Wales, and in Queensland are associated with the late Paleozoic or early Mesozoic position of segments of the applicable seismotectonomagmatic belts. The Cenozoic copper deposits in the north island of New Zealand and in the chain of islands north of Australia and New Zealand are associated with the present (Cenozoic) position of the Bengal-to-Kermadecs, Kermadecs-to-Galapagos, and Kermadecs-to-Bouvet segments of the equatorial and of one meridional seismotectonomagmatic belt (Chapter 4).

Present Types of Copper Deposits as a Function of the Earth's Behavior.

When the dynamic behavior of the Earth, during the past 4.6 b. y., is considered on a global scale, it is not too surprising that this behavior has produced a great variety of types of copper deposits. If the Earth's behavior has been motivated by a long-lived, deep-seated driving mechanism, such as that provided by the Tectonospheric Earth Model, then it follows that certain constraints would have operated to minimize the total number of types of copper deposits found today.

Copper is concentrated in many different kinds of geological environments. Much copper occurs in porphyries where the mineralization is associated with intrusive igneous rocks of acid to intermediate composition. More-or-less even disseminations of copper permeate large volumes of rock to produce large tonnages of low-grade ore. Some deposits resulted from the enrichment of lean mineralizations. Where natural solvents (such as pyrite) were present, weathering processes removed copper from the oxidized zone and precipitated it at or just below the water table, thereby producing a rich cover of copper ore. Some of the highest-grade ores come from vein and contact-metamorphic deposits wherein copper minerals fill fractures, pores, and vugs and replace part of the surrounding rock. Tonnages of this type are seldom large except at a few locations such as Butte, Montana. Extensive stratiform bodies form large tonnages of copper in sedimentary rocks such as sandstone, shale, and conglomerate. The evolution of the ore appears to have been controlled by the sedimentary features as well as by later structures. The deposits in central Africa fall into this category. In old Precambrian shield areas, copper is associated with Ni in ultrabasic rocks. In the Lake Superior area, the copper occurs in lava flows and associated conglomerates, sandstones, and shales.

When copper deposits are analyzed on a global basis, various classification schemes may be used to show what percentages of copper deposits fall into each type. One of these may be summarized:

a. Porphyry copper deposits with molybdenum 48%

b. Stratiform deposits disseminated in sediments and poor in sulfur and iron 27%

c. Massive pyrite deposits associated with volcanic rocks 10%

d. Deposits of enargite 4%

e. Deposits of iron oxide associated with basic intrusive rocks 2%

f. Deposits of native copper associated with basic lavas 2%

g. Deposits with abundant nickel associated with basic and ultrabasic intrusives. 2%

h. Deposits with arseno-pyrite or pyrrhotite and gold 1%

i. Stratiform impregnations associated with volcanic rocks poor in pyrite. 1%

j. Deposits with tennantite containing germanium 1%

k. Cupriferous tin deposits 1%

1. Others

From this, it may be seen that three-fourths of all the copper is now concentrated within just two types of deposits: (1) porphyry copper with molybdenum, and (2) disseminated stratiform deposits. This suggests that a long-lived, deep-seated mechanism has controlled the origin, evolution, and present characteristics of the Earth's copper deposits. It is the author's thesis that this mechanism is that provided by the Tectonospheric Earth Model (Chapter 2: page 14).

This suggests that the Tectonospheric Earth Model should provide a supplementary tool in the exploration for copper deposits. Before describing how the Tectonospheric Earth Model may be used for exploration work, it is well to summarize how copper deposits are formed, as interpreted by this model.

The Formation of Copper Deposits.

Depending upon the specific system used for classifying the applicable formation processes, copper deposits may be divided into many kinds. Alternatively, any copper deposit may be considered to result from a combination of processes. For any Cu deposit, these processes may be expressed schematically:

$$\sum_{i=1}^{n} a_i A_i$$ where $i = 1, 2, 3, --- n$; A_i are the various processes that can contribute to the formation of copper deposits; and a_i are constants, any of which may be zero, provided that

$$\sum_{i=1}^{n} a_i = 1.$$

A few of the processes, A_i, may be summarized:

A_1 = Magmatic Processes. During the crystallization of magmas (rock melts) within the Earth's tectonosphere, economically important quantities of chromite ($FeCr_2O_4$) and compounds of Pt and the platinum group of metals may separate from an intrusive (or extrusive) mass. Examples of this process are found in the chromite deposits in the Bushveld Complex of South Africa. There, hundreds of millions of tons of chromite are found, as well as enough Pt to supply the world's needs for hundreds of years. Many copper deposits owe their existence to magmatic process, e. g., the Arizona-Sonora deposits.

A_2 = Hydrothermal Processes. Many magmas contain several percent of water, some of which is retained in the crystallization of hydrous silicate minerals. Most of the water, however, escapes volatilely during the crystallization process. If the magma contains at least a few ppt (parts per thousand) of chloride, then the escaping volatiles are highly saline and are capable of providing sinks for metals such as Pb, Zn, and Mn, causing these metals to be extracted from the magma. Part of the sulfur constituent of the magma also escapes during this process. Upon cooling, some of the metals, such as Fe, Sn, and W are precipitated as oxides. Others, such as Cu, Pb, Zn, Ag, As, and Sb, are precipitated as sulfides. Still others, such as Au and Ag, are precipitated as pure metals. In this process, the time interval between the emplacement of a magma and the formation of the associated ore deposits is usually less than 10^5 years. Shallow ground water can penetrate into hot intrusives and thereby participate in the process of metal extraction and ore deposition. In oceanic areas, seawater may intrude into hot igneous rocks at or below the ocean floor. Examples include: (1) the extensive stratiform Cu, Pb, and Zn deposits of Japan and other locations designated as Kuroko-type deposits; (2) the Cu deposits of Cyprus; and (3) the very extensive sediments rich in Fe and Mn associated with some mid-ocean ridges and rises. Examples in which the deposits are less-clearly associated with magmas include: (1) the large Pb-Zn deposits of midcontinental United States, and other geologically similar provinces; and (2) the extensive uranium deposits of the Colorado Plateau of the United States.

A_3 = Residues of Weathering. During the process of chemical weathering, the alkaline substances and silica are normally removed preferentially from silicate substances. The resulting residues are usually enriched in Al and Fe. Further enrichment produces substances consisting largely of hydrated aluminum oxides (bauxites). Flowing water often contributes to the weathering process, with the result that many river and beach sands can be mined for Au, Pt, cassiterite (SnO_2), Ti, and Zr. Examples include the gold placers of the western United States and of South Africa. The latter, with the associated uranium, are generally considered to be ancient placer that have undergone a certain amount of remobilization during post-depositional burial.

A_4 = Sedimentary Processes. Most of the Earth's largest iron deposits were precipitated from surface waters. Examples include the Fe deposits of North America (Lake Superior), Australia, South Africa, the U. S. S. R., and South America. Most of these were precipitated during a relatively short time, about 2 b. y. ago. Other examples include the concentrations of Co, Cu, and Ni in ocean-floor nodules; calcium phosphates and uranium deposited near upwelling areas within the oceans; and uranium in organic-rich shales, such as the Chattanooga black shales of the eastern United States.

A_i (i = 5, 6, 7, --- n) = Other Processes. The origins

of some copper deposits have been influenced by processes other than those represented by A_1, A_2, A_3, and A_4. These need not be listed.

Several examples may be cited within the above scheme:

<u>The Arizona-Sonora Copper Deposits</u> may be expressed schematically as

$$\sum_{i=1}^{n} a_i A_i \quad \text{where } a_1 = a_2 = 0.5\text{, all other } a_i = 0\text{, and}$$

A_1 = magmatic processes during the Precambrian; and
A_2 = hydrothermal processes during the Phanerozoic.

<u>The Kuroko Type Copper Deposits</u> may be expressed schematically as

$$\sum_{i=1}^{n} a_i A_i \quad \text{where } a_2 = 1\text{, all other } a_i = 0\text{, and}$$

A_2 = hydrothermal processes.

<u>The Copper Concentrations within Ocean-Floor Nodules</u> may be expressed schematically as

$$\sum_{i=1}^{n} a_i A_i \quad \text{where } a_4 = 1.0\text{, all other } a_i = 0\text{, and}$$

A_4 = sedimentary processes.

The processes that served to produce each of the Earth's copper deposits may be expressed schematically in a manner similar to the above examples. It is not necessary, for this analysis, that others be listed. Suffice it here to say merely that this scheme, expanded into a spatio-temporal framework, serves as the basis for integrating the several applicable processes that have operated, at various times during the past 4.6 b. y., to produce today's copper deposits.

Chapter 12

THE TECTONOSPHERIC EARTH MODEL AS A SUPPLEMENTARY TOOL IN THE EXPLORATION FOR COPPER DEPOSITS

Most easy-to-find copper deposits in the populated regions of the world have been discovered. New discoveries are becoming increasingly more difficult and require more time and money. This suggests that new methods must be developed to discover concealed copper deposits. The Tectonospheric Earth Model provides one such method. It does this, primarily, by defining a single mechanism for all the Earth's geometrical, mechanical, thermal, and chemical behavior that has occurred during the 4.6 b. y. that the Earth is believed to have been in existence (Chapter 2: page 14). The model-defined activity includes (1) Archean protocratonic development; (2) Proterozoic non-rigid cratonic, plate, and block behavior; (3) Phanerozoic rigid cratonic, plate, and block behavior; and (4) the orogenic and related seismotectonomagmatic behavior associated with (1), (2), and (3).

The Basis for Using the Tectonospheric Earth Model as a Supplementary Tool in the Exploration for Copper Deposits.

The earlier chapters have considered some of the geometrical, mechanical, thermal, and chemical aspects of the Tectonospheric Earth Model that may be used to derive the origin, evolution, and present characteristics of the surficial manifestations of the Earth's internal behavior. These surficial manifestations include: the orogenic-cratonic structure of the continents; the Earth's deep seismicity; global patterns of terrestrial heat flow and other geothermal activity; geosynclinal evolution and mountain building;

global seismotectonomagmatic and mineralization belts; the Earth's gravity field; geomagnetism and polarity reversals; seafloor spreading and other forms of crustal rifting; polar wandering and continental drift; mantle-plume development; and the general omniductive, plate-tectonic behavior of the Earth (Tatsch, 1972a: pages 148-197).

One of the special features of the Tectonospheric Earth Model is that it permits a worldwide geological synthesis extending over a large spatio-temporal frame. Also, this model permits an analysis of subsurface conditions in areas where local surface structure may not be a clue to the presence of the deeper structure, e. g., where Tertiary rocks at the surface might be separated from the underlying Cretaceous strata by angular unconformities.

As stated earlier, it has been known for some time that certain copper provinces appear to be associated with plate-tectonic boundaries and with other tectonomagmatic, seismotectonic, and omniductive features of the Earth. The Tectonospheric Earth Model corollary that permits an analysis of this association may be summarized: the origin, evolution, migration, and accumulation of copper provinces are related, in a predictable manner, to the seismotectonomagmatic belts that have been formed (or modified, in some cases) during the existence of the Earth.

When applied to copper deposits, the model permits one to make working-hypothesis type predictions about the salient relationships that exist: (1) between copper deposits and geologic structure; (2) between specific types of copper deposits and the controlling thermal and stress patterns of the Earth; and (3) among the other existing relationships that are not predictable from other models when considering large areas over long periods of time.

The Tectonospheric Earth Model is used also to provide the basic "fixed" frame of reference for making predictions regarding the geometrical, mechanical, thermal, and chemical behavior of the

Earth's upper 1000 km during the past 4.6 b. y. in those cases where other applicable models and concepts are, non-committal, or non-existent.

<u>The Outputs of the Model</u>.

This book is based on the first-order, unadjusted predictions of the Tectonospheric Earth Model (Tatsch, 1972a: chapter 20), using the above assumptions as the basis for inputs to the model. No attempt is made in this book to use the results of the higher-order capabilities of the model. Rather, the effort of this analysis is directed toward a simple answer to the question: What is the probability for the existence of economically attractive copper deposits in various parts of the world?

The basic outputs of the model, under the above simplifications, consist essentially of predictions based on: (1) the local geologic structure as the control mechanism for the present accumulations of copper deposits within various areas of the world; (2) the paths of copper from the ultimate source to form the present deposits; (3) the applicable processes in, and the properties of, the host rocks associated with the present deposits; (4) the origin, evolution, migration, and role of fluids in connection with the development of present deposits; (5) the origin, evolution, migration, and role of heat in connection with the development of present deposits; and (6) the origin, evolution, migration, and role of stress patterns in connection with the devlopment of the present deposits.

From the above, it may be seen that the modeling concept used in the Tectonospheric Earth Model methodology differs significantly in two salient respects from conventional modeling concepts: (1) the Tectonospheric Earth Model concept is long-lived, deep-seated, and realistic in that it imposes "nature" on the model (rather than the converse, which is seen so often today in mathematically-tidy but otherwise hopelessly-simplified models); and (2) it uses mathematics as a flexible tool (rather than as a rigid framework that renders most mathematically-tidy models fairly useless in real-world situations where nature's actual problems face us in connection with the origin, evolution, and present characteristics of copper deposits).

Recommendations.

Two specific recommendations may be made regarding particular applications for using the Tectonospheric Earth Model as a supplementary tool in the exploration for copper deposits:

a. That the Tectonospheric Earth Model be used in its lowest order of resolution (i. e., in its first order, with a nominal resolution of 1571 km) in connection with initial analyses for Cu deposits, particularly in unknown and previously unexplored areas within the applicable segments of the twelve surficial arcs associated with the seismotectonomagmatic belts of the model described in Chapter 4.

b. That the model be used in its higher orders of resolution when making more advanced determinations in known and previously explored areas where greater resolution is desired. For example, in its second, third, and fourth orders, the nominal resolutions are 157.1, 15.71, and 1.571 km, respectively. By using certain fractional orders, any desired nominal resolution may be obtained, ranging from 1571 km to zero. Thus, an order of 1.5 produces a nominal resolution of about 1100 km, an order of 2.5 produces a nominal resolution of about 110 km, an order of 3.5 produces a nominal resolution of about 11 km, an order of 4.5 produces a resolution of about 1.1 km, and an order of 5.5 produces a resolution of about 110 m.

In each real-world case, the actual resolution for any given area or project is optimized on the basis of the amount of time and funding available for that area or project, because the resolution obtained is a function of both the time and funding available. Thus, if the desired resolution and time are specified for a given project or area, then the amount of funding required within these constraints for that area or project can be computed.

Finally, it is felt that this approach is particularly valuable as a supplementary tool in the exploration for Cu deposits within those areas that contain numerous heterogeneities. Space does not permit a discussion here of the details regarding the handling of the various geometrical, mechanical, thermal, and chemical heterogeneities encountered within the Earth's tectonosphere. These have been outlined briefly in Tatsch (1972a), and the author feels that one of the strengths of the Tectonospheric Earth Model is its ability to predict the variations within these heterogeneities in all parts of the tectonosphere during the 4.6 b. y. that the Earth is believed to have been in existence.

The author feels, in conclusion, that the Tectonospheric Earth Model provides a means of using the scant paucity of orderliness within the Earth's copper-distribution patterns to increase the so-called "signal-to-noise ratio" in these patterns. This, in turn, permits the model to predict the locations of undiscovered copper deposits.

REFERENCES

NOTE: Abbreviations used in this section are explained at the end of the section.

Abbott, D. M., Jr., and Tautwein, C. M., 1975. Recognition of seven periods of Precambrian folding in the Colorado Front Range. Geol. Soc. Amer. Abstr. Prog., 7, 585.

Adaima, Sh. A., 1975. Plate tectonics and the development of the Alpine system: Discussion. Geol. Soc. Amer. Bull., 86, 719-720.

Ahmad, F., 1973. Have there been major changes in the Earth's axis of rotation through time? In: D. H. Tarling and S. K. Runcorn (Editors), Implications of Continental Drift to the Earth Sciences, vol. 1. Academic, New York, pp. 487-501.

Ahrens, L. H. (Editor), 1968. Origin and Distribution of the Elements. Pergamon, New York, 1178 pp.

Ahrens, T. J., 1973. Petrologic properties of the upper 670 km of the Earth's mantle: geophysical implications. Phys. Earth Planet. Int., 7, 167-186.

Aki, K., 1973. Scattering of P-waves under the Montana Lasa. J. Geophys. Res., 78, 1334-1346.

Alexandrov, E. A., 1955. Contribution to studies of origin of Precambrian banded iron ores. Econ. Geol. 50, 459-368.

Allan, D. W., Thompson, W. B., and Wesss, N. O., 1967. Convection in the Earth's mantle. In: S. K. Runcorn (Editor), Mantles of the Earth and Terrestrial Planets. Interscience, New York, pp. 507-512.

Alldredge, L. R., and Hurwitz, L., 1964. Radial dipoles as

the source of the Earth's main magnetic field. J. Geophys. Res., 69, 2631-2640.

Allegre, C. J., Javoy, M., and Michard, G., 1968. Etude de la distribution et de l'abondance des elements de transition dans l'ecorce terrestre, compares a celles des terres rares. In: L. H. Ahrens (Editor), Origin and Distribution of the Elements. Pergamon, New York, pp. 913-928.

Allen, C. R., 1962. Circum-Pacific faulting in the Philippine-Taiwan region. J. Geophys. Res., 67, 4795-4812.

Allsop, H. L., 1961. Rb-Sr age measurements on total rock and separated-mineral fractions from the Old Granite of the Central Transvaal. J. Geophys. Res., 66, 1499-1508.

Alvarez, W., 1973. The application of plate tectonics to the Mediterranean region. In: D. H. Tarling and S. K. Runcorn (Editors), Implications of Continental Drift to the Earth Sciences, vol 2. Academic, New York, pp. 893-908.

Amstutz, G. C., and Bernard, A. J. (Editors), 1973. Ores in Sediments. Springer-Verlag, Berlin, 350 pp.

Anderle, R. J., and Smith, S. J., 1968. Observation of 27th and 28th order gravity coefficients based on Doppler observations. J. Astronaut. Sci., 15, 1-4.

Anderson, D. L., 1967. Phase change in the upper mantle. Science, 157, 1165-1168.

Anderson, D. L., 1969. Geophysical evidence on the petrology of the mantle. Geol. Soc. Amer. Abstr. Prog., 7, 2.

Anderson, O. L., 1972. Patterns in elastic constants of minerals important to geophysics. In: E. C. Robertson (Editor), The Nature of the Solid Earth. McGraw-Hill, New York, pp. 575-613.

Anderson, T. A., 1975. Carboniferous subduction complex in the Harz Mountains, Germany. Geol. Soc. Amer. Bull., 86, 77-82.

Anderson, T. H., and Silver, L. T., 1974. Late Cretaceous plutonism in Sonora, Mexico, and its relationship to Circum-

Pacific magmatism. Geol. Soc. Amer. Abstr. Prog., 6, 484.

Anderson, D. L., Sammis, C., and Jordan, T., 1972. Composition of the mantle and core. In: E. C. Robertson (Editor), The Nature of the Solid Earth. McGraw-Hill, New York, pp. 41-66.

Anderson, R. N., Clague, D. A., Klitgord, K. D., Marshall, M., and Nishimori, R. K., 1975. Magnetic and petrologic variations along the Galapagos spreading center and their relation to the Galapagos melting anomaly. Geol. Soc. Amer. Bull., 86, 683-694.

Annels, A. E., 1974. Some aspects of the stratiform ore deposits of the Zambian Copperbelt and their genetic significance. In: P. Bartholome (Editor) (q. v.), pp. 235-254.

Argand, E., 1916. Sur l'arc des Alpes occidentales. Eclogae Geologicae Helvetiae, 16, 179-182.

Armstrong, R. L., 1968. A model for the evolution of strontium and lead isotopes in a dynamic Earth. Rev. Geophys., 6, 175-199.

Armstrong, R. L., 1971. Isotopic and chemical constraints on models of magma genesis in volcanic arcs. Earth Planet. Sci. Ltr., 12, 137-142.

Armstrong, R. L., and Cooper, J. A., 1971. Lead isotopes in island arcs. Bull. Volcanol., ser. 2, 35, 27-63.

Armstrong, F. E., and Heemstra, R. J., 1972. Radiometrics proposed for exploration. Oil Gas J., Jun. 5, 1972, pp. 88-91, 97.

Arth, J. G., and Hanson, G. N., 1975. Geochemistry and origin of the early Precambrian crust of northeastern Minnesota. Geochim. Cosmochim. Acta, 39, 325-362.

Artyushkov, E. V., 1974. Can the Earth's crust be in a state of isostasy? J. Geophys. Res., 79, 741-752.

Assanov, M. A., Essenov, Ch. E., Mouraline, N. N., and Seifoulline, S. Ch., 1974. Geologie et metallogenie des gisements stratiformes de cuivre (gres cuivreux) du Kazakhstan central d'apres l'example du Djezkazgan. In: P. Bartolome (Editor) (q. v.), pp. 339-352.

Athavale, R. N., 1973. Inferences from recent Indian paleomagnetic results about the northern margin of the Indian plate and the tectonic evolution of the Himalayas. In: D. H. Tarling and S. K. Runcorn (Editors), Implications of Continental Drift to the Earth Sciences, vol. 1. Academic, New York, pp. 117-130.

Atwater, T., 1973. Studies of sea-floor spreading and plate movements in the Pacific basin. In: D. H. Tarling and S. K. Runcorn (Editors), Implications of Continental Drift to the Earth Sciences, vol. 1. Academic, New York, pp. 213-217.

Aubouin, J., 1965. Geosynclines. Elsevier, Amsterdam, 355 pp.

Audley-Charles, M. G., 1975. The Sumba fracture: A major discontinuity between eastern and western Indonesia. Tectonophys., 26, 213-228.

Audley-Charles, M. G., Carter, D. J., and Milsom, J. S., 1972. Tectonic development of eastern Indonesia in relation to Gondwanaland dispersal. Nature Phys. Sci., 239, 35-39.

Avdeiko, G. P., 1971. Evolution of geosynclines on Kamchatka. Pac. Geol., 3, 1-14.

Ayres, D. E., 1972. Genesis of iron-bearing minerals in banded iron formation mesobands in the Dales Gorge Member, Hammersley Group, Western Australia. Econ. Geol., 67, 1214-1233.

Azhgirei, G. D., 1968. Folding and mountain building (orogenesis). In: Mistik (Editor), Orogenic Belts. Academia, Prague, pp. 287-289.

Baer, A. J. (Editor), 1970. Basins and Geosynclines of the Canadian Shield (GSC Spl Ppr 70-40). Geological Survey of Canada, Ottawa, 265 pp.

Baer, A. J., 1975. The Grenville belt. Mtg. Roy. Soc. Lond., Mar. 13-14.

Bain, G. W., 1968. Syngenesis and epigenesis of ores in layered rocks. In: M. Stemprok (Editor), Endogenous Ore Deposits. Academia, Prague, pp. 119=136.

Baird, A. K., Morton, D. M., Woodford, A. O., and Baird, K. W., 1974. Transverse Range Province: A unique structural-petrochemical belt across the San Andreas fault system. Geol. Soc. Amer. Bull., 85, 163-174.

Baker, B. H., Mohr, P. A., and Williams, L. A. J., 1972. Geology of the Eastern Rift System of Africa (GSA Spl Ppr 136). Geological Society of America, Boulder, 67 pp.

Balakina, L. M., Misharina, L. I., Shirkova, E. I., Vredenskaya, A. V., 1969. The field of electrical stresses associated with earthquakes. In: P. J. Hart (Editor), The Earth's Crust and Upper Mantle. Amer. Geophys. Union, Washington, pp. 166-171.

Bamford, R. W., 1972. The Mount Fubian (Ok Tedi) porphyry copper deposit, Territory of Papua and New Guinea. Econ. Geol., 67, 1019-1033.

Barazangi, M., and Dorman, J., 1969. World seismicity map of ESSA Coast and Geodetic Survey epicenter data for 1961-1967. Bull. Seis. Soc. Amer., 59, 369-380.

Barazangi, M., Pennington, W., and Isacks, B., 1975. Global study of seismic wave attenuation in the upper mantle behind island arcs using pP waves. J. Geophys. Res., 80, 1079-1092.

Barth, T. F. W., 1968. The geochemical evolution of continental rocks: A model. In: L. H. Ahrens (Editor), Origin and Distribution of the Elements. Pergamon, New York, pp. 587-597.

Bartholome, P., 1974a. On the diagenetic formation of ores in sedimentary beds, with special reference to Kamoto, Shaba, Zaire. In: P. Bartholome (Editor), (q. v.), pp. 203-213.

Bartholome, P.,(Editor), 1974b. Gisements Stratiformes et Provinces Cupriferes. Soc. Geol. Belg., Liege, 427 pp.

Bartura, Y., and Wurzburger, U., 1974. The Timna copper

deposit. In: P. Bartholome (Editor) (q. v.), pp. 277-285.

Bass, M. N., 1970. Variation of ocean basalts with spreading rate. Trans. Amer. Geophys. Union, 51, 762.

Bateman, A. M., 1950. Economic Mineral Deposits, 2nd ed. Wiley, New York, 916 pp.

Bateman, A. M., 1951. The Formation of Mineral Deposits. Wiley, New York, 371 pp.

Bauchau, C., 1971. Easai de typologie quantitative des gisements de plomb et de zinc avec la repartition de l'argent. Bureau de Recherches Geologiques et Minieres, Orleans, 118 pp.

Bauer, H. L., Breitrick, R. A., Cooper, J. J. and Anderson, 1966. Porphyry copper deposits in the Robinson mining district, Nevada. In: Titley and Hicks (Editors) (q. v.), pp. 233-244.

Bell, R. T., 1970. The Hurwitz Group - A prototype for deposition of metastable cratons. In: A. J. Baer (Editor), Basins and Geosynclines of the Canadian Shield. Geol. Survey of Canada, Ottawa, pp. 159-170.

Beloussov, V. V., 1962. Basic Problems in Geotectonics. McGraw-Hill, New York, 816 pp.

Beloussov, V. V., 1968. Some general aspects of development of the tectonosphere. In: K. Benes (Editor), Upper Mantle (Geological Processes). Academia, Prague, pp. 9-17.

Beloussov, V. V., 1969a. Continental rifts. In: P. J. Hart (Editor), The Earth's Crust and Upper Mantle. Amer. Geophys. Union, Washington, pp. 539-544.

Beloussov, V. V., 1969b. Interrelations between the Earth's crust and upper mantle. In: P. J. Hart (Editor), The Earth's Crust and Upper Mantle. Amer. Geophys. Union, Washington, pp. 698-712.

Beloussov, V. V., 1970. Against the hypothesis of ocean floor spreading. Tectonophys., 9, 489-511.

Beloussov, V. V., 1971. On possible forms of relationships between magmatism and tectonogenesis. J. Geol. Soc., 127, 57-68.

Beloussov, V. V., 1972. Important trends in the evolution of continents. Internat. Geol. Cong. Abstr., 24, 65.

Benerjee, P. K., and Ghosh, S., 1972. Correlation of Precambrian ore provinces of East Africa, India, and West Australia. Econ. Geol., 67, 55-62.

Benes, K. (Editor), 1968. Upper Mantle (Geological Processes). Academia, Prague, 260 pp.

Benioff, H., 1959. Circum-Pacific tectonics. Publs. Domin. Obs. Ottawa, 20, 395-402.

Bennett, G. T., Clowes, R. M., and Ellis, R. M., 1975. A seismic refraction survey along the southern Rocky Mountain Trench, Canada. Bull. Seis. Soc. Amer., 65, 37-54.

Benson, W. N., 1926. The tectonic conditions accompanying the intrusion of basic and ultrabasic igneous rocks. Nat. Acad. Sci. Mem., 19, 1-90.

Berg, E., and Sutton, G. H., 1975. Dynamic interaction of seismic and volcanic activity of the Nazca plate edges. Phys. Earth Planet. Int., 9, 175-182.

Berg, E., Sutton, G. H., and Walker, D. A., 1975. Dynamic interaction of seismic activity along rising and sinking edges of Indian and Pacific Ocean plate boundaries. Geol. Soc. Amer. Abstr. Prog., 7, 395.

Berge, J. W., 1972. Physical and chemical factors in the formation of marine apatite. Econ. Geol., 67, 824-827.

Berkland, J. D., 1972. Neogene "frozen" subduction zone in the coast ranges of northern California. Internat. Geol. Cong. Abstr., 24, 65.

Bernard, J. H., 1970. Gradual decrease of the spatial relation between the Variscan granitoid plutons and the persistent ore-forming hydrothermal mineral associations of the Bohemian massif. In: Z. Pouba and M. Stemprok (Editors), Problems of Hydrothermal Ore Deposits. Schweitzerbartsche V., Stuttgart, pp. 58-60.

Bernard, A. J., and Soler, E., 1974. Apercu sur la province pyriteuse Sud-Iberique. In: P. Bartholome (Editor) (q. v.), pp. 287-315.

Best, M. G., and Brimhall, W. H., 1974. Late Cenozoic alkalic basaltic magmas in the western Colorado plateaus and the Basin and Range transition zone, U. S. A., and their bearing on mantle dynamics. Geol. Soc. Amer. Bull., 85, 1677-1690.

Bickford, M. E., and Van Schmus, W. R., 1973. Possible middle and late Precambrian igneous arcs in the mid-continent region of North America. Geol. Soc. Amer. Abstr. Prog., 5, 300.

Binda, P. L., and Mulgrew, J. R., 1974. Stratigraphy of copper occurrences in the Zambian Copperbelt. In: P. Bartholome (Editor) (q. v.), pp. 215-233.

Biq, C., 1960. Circumpacific tectonics in Taiwan. Rept. Internat. Geol. Cong., 21st Cong., 18, 203-214.

Bird, J. F., 1970. General concepts of orogenesis in terms of lithospheric plate tectonics. Geol. Soc. Amer. Abstr. Prog., 2, 733-734.

Black, L. P., and Richards, J. R., 1972. Isotopic composition and possible genesis of ore leads in northeastern Queensland, Australia. Econ. Geol. 67, 1168-1179.

Black, L. P., Gale, N. H., Moorbath, S., Pankhurst, R. J. and McGregor, V. R., 1971. Isotopic dating of very early Precambrian amphibolite facies gneisses from the Godthaab district, West Greenland. Earth Planet. Sci. Ltr., 12, 245-259.

Blackett, P. M. S., Bullard, E. C., and Runcorn, S. K., (Editors), 1965. A symposium on continental drift. Phil. Trans. Roy. Soc., Ser. A., 258; 1

Blatt, H., Middleton, G., and Murray, R., 1972. Origin of Sedimentary Rocks. Prentice-Hall, Englewood Cliffs, N. J., 634 pp.

Bloom, H., 1966. Geochemical exploration as applied to copper-molybdenum deposits. In: Titley and Hicks (Editors) (q. v.), pp. 111-119.

Bogdanov, Y. V., and Kutyrev, E. I., 1972. Geological conditions of the distribution of stratified copper and lead-zinc deposits of the USSR territory. Internat. Geol. Cong. Abstr., 24, 112-113.

Bohmer, M., and Stohl, J., 1968. Laws of locations of development of endogenous mineralization in neovolcanites of central Slovakia. In: M. Stemprok (Editor), Endogenous Ore Deposits. Academia, Prague, pp. 241-253.

Boldt, J. R., 1967. The Winning of Nickel. Van Nostrand, New York, 487 pp.

Boles, J. R., and Coombs, D. S., 1975. Mineral reactions in zeolitic Triassic tuff, Hokonui Hills, New Zealand. Geol. Soc. Amer. Bull., 86, 163-173.

Bonatti, E., Fisher, D. E., Joensuu, O., Rydell, H. S., and Beyth, M., 1972. Iron-manganese-barium deposit from the northern Afar rift (Ethiopia). Econ. Geol., 67, 717-730.

Bonini, W. E., Loomis, T. P., and Robertson, J. D., 1973. Gravity anomalies, untramafic intrusions, and the tectonics of the region around the Strait of Gibraltar. J. Geophys. Res., 78, 1372-1382.

Bonisevich, A. S., Riabinkin, L. A., and Turchanenko, N. T., 1971. Seismic investigation of deep, complex structures. Eighth World Petroleum Conference, vol. 3. Applied Science Publishers, Ltd., London, pp. 3-13.

Borrello, A. J., 1972. The Precordillera as a type of geosyncline in Argentina. Internat. Geol. Cong. Abstr., 24, 67.

Bostrom, K., Farquharson, B., and Eyl, W., 1972. Submarine hot springs as a source of active ridge sediments. Chem. Geol., 10, 189-198.

Bott, M. H. P., 1973a. The evolution of the Atlantic north of the Faeroe Islands. In: D. H. Tarling and S. K. Runcorn (Editors), Implications of Continental Drift to the Earth Sciences,

Bott, M. H. P., 1973b. Shelf subsidence in relation to the evolution of young continental margins. In: D. H. Tarling and S. K. Runcorn (Editors), Implications of Continental Drift to the Earth Sciences, vol. 2. Academic, New York, pp. 675-683.

Bottinga, Y., 1974. Thermal aspects of sea-floor spreading, and the nature of the suboceanic lithosphere. Tectonophys., 21, 15-38.

Boucot, A. J., 1972. Silurian of Venezuela. In: W. B. N. Berry and A. J. Boucot (Editors), Correlation of South American Silurian Rocks (GSA Spl Ppr 133). Geol. Soc. Amer., Washington, 59 pp.

Bowen, R. L., 1973. Cenozoic dimensional changes of continental and oceanic crusts. Geol. Soc. Amer. Abstr. Prog., 5, 248.

Bowen, R. L., 1974. The enigma of Late Paleozoic orogeny in southeastern North America. Geol. Soc. Amer. Abstr. Prog., 6, 336-337.

Bowes, D. R., 1975. Tectonics in the Baltic Shield in the period 2000-1500 million years. Mtg. Roy. Soc. Lond., Mar. 13-14.

Bowin, C., 1974. Migration of a pattern of plate motion. Earth Planet. Sci. Ltr., 21, 400-404.

Boyle, R. W., and Davis, J. L., 1973. Banded iron formations. Geochim. Cosmochim. Acta, 37, 1389-1390.

Braitsch, O., 1971. Salt Deposits, Their Origin and Composition. Springer-Verlag, New York, 297 pp.

Brant, A. A., 1966. Geophysics in the exploration for Arizona porphyry coppers. In: Titley and Hicks (Editors) (q. v.), pp. 87-110.

Brewer, M. S., and Lippolt, H. J., 1974. Petrogenesis of basement rocks of the upper Rhine region elucicated by rubidium-strontium systematics. Contr. Mineral. Petrol., 45, 123-141.

Briden, J. C., Piper, J. D. A., Henthorn, D. I., and Rex, D. C., 1971. 15 Ann. Rpt. Res. Inst. Af. Geol., Leeds, 46-50.

Brobst, D. A., and Pratt, W. P. (Editors), 1973. United States Mineral Resources. Government Printing Office, Washington, 722 pp.

Brock, B. B., 1972. A Global Approach to Geology: The Background of a Mineral Exploration Strategy Based on Significant Form in the Patterning of the Earth's Crust. Balkema, Cape Town, 365 pp.

Brock, A., 1973. Precambrian polar wander and continental drift of Africa and North America: Some comments. In: D. H. Tarling and S. K. Runcorn (Editors), Implications of Continental Drift to the Earth Sciences, vol. 1. Academic, New York, pp. 33-36.

Brown, A. C., 1971. Zoning in the White Pine copper deposit, Ontonagon County, Michigan. Econ. Geol. 66, 543-573.

Brown, J. S., 1973. Sulfur isotopes of Precambrian sulfates and sulfides in the Grenville of New York and Ontario. Econ. Geol., 68, 362-370.

Brown, A. C., 1974a. The copper province of northern Michigan, U. S. A. In: P. Bartholome (Editor) (q. v.), pp. 317-330.

Brown, B. W., 1974b. Sediment accumulation, sea-floor spreading, and eustacy: Discussion. Geol. Soc. Amer. Bull., 85, 831.

Browne, P. R. L., and Lovering, J. F., 1973. Composition of sphalerites from the Broadlands (New Zealand) geothermal field and their significance to sphalerite geothermometry and geobarometry. Econ. Geol., 68, 381-387.

Brune, J. N., and Dorman, J., 1963. Seismic waves and Earth structure in the Canadian Shield. Bull. Seis. Soc. Amer., 53, 167-210.

Bullard, E. C., 1949. The magnetic field within the Earth. Proc. Roy. Soc., A, 197, 433.

Bullard, E. C., 1956. Edmund Halley (1656-1742). Endeavor, 15, 189.

Bullard, E. C., 1964. Continental drift. Quart. J. Geol. Soc. Lond., 120, 1

Bullard, E. C., Everett, J. E., and Smith, A. S., 1965. The fit of the continents around the Atlantic. Phil. Trans. Roy. Soc. Lond., A, 258, 41-51.

Bullen, K. E., 1974a. Introductory remarks on standard Earth model. Phys. Earth Planet. Int., 9, 1-3.

Bullen, K. E., 1974b. Standard Earth model requirements in respect of density and rigidity in the inner core. Phys. Earth Planet. Int., 9, 41-44.

Burchfiel, B. C., and Davis, G. A., 1968. Two-sided nature of the Cordilleran orogen and its tectonic implications. In: M. Mistik (Editor), Orogenic Belts. Academia, Prague, pp. 175-184.

Burchfiel, B. C., and Davis, G. A., 1972. Structural framework and evolution of the southern part of the Cordilleran orogen, western United States. Amer. J. Sci., 272, 97-118.

Burk, C. A., 1972. Global tectonics and world resources. Amer. Assoc. Petrol. Geol. Bull., 56, 196-202.

Burk, C. A., 1975. Sea of Okhotsk: Thirteenth cruise of the Dmitry Mendeleev. Geology, 3, 141-144.

Burke, K., and Dewey, J., 1973. An outline of Precambrian plate developments. In: D. H. Tarling and S. K. Runcorn (Editors), Implications of Continental Drift to the Earth Sciences, vol. 2. Academic, New York, pp. 1035-1045.

Burke, K., and Whiteman, A. J., 1973. Uplift, rifting, and the breakup of Africa. In: D. H. Tarling and S. K. Runcorn (Editors), Implications of Continental Drift to the Earth Sciences, vol. 2. Academic, New York, pp. 735-755.

Burnham, C. W., 1967. Hydrothermal fluids at the magmatic stage. In: H. L. Barnes (Editor), Geochemistry of Hydrothermal Ore Deposits. Hllt, Rinehart, and Winston, New York, pp. 166-235.

Burrett, C. F., 1974. Plate tectonics and the fusion of Asia. Earth Planet. Sci. Ltr., 21, 181-189.

Burtman, V. S., 1972. Development of geosynclinal folding.

Geotectonics, *2*, 79-82.

Butler, B. S., 1942. Some interrelations of structure, mineralogy, and association with intrusive bodies in ore deposits. In: W. H. Newhouse (Editor), *Ore Deposits Related to Structural Features*. Princeton Univ. Press, Princeton, pp. 3-5.

Cabre, R., 1975. Geodynamics in the eastern Pacific and the western Americas. *Phys. Earth Planet. Int.*, *9*, 169-173.

Cady, W. M., 1973. The earmarks of sudbuction. In: D. H. Tarling and S. K. Runcorn (Editors), *Implications of Continental Drift to the Earth Sciences*, vol. 2. Academic, New York, pp. 955-961.

Cady, W. M., 1974. Toward a rationale in geotectonics. *Trans. Amer. Geophys. Union*, *55*, 438.

Cahen, L., 1974. Geological background to the copper-bearing strata of southern Shaba (Zaire). In: P. Bartholome (Editor) (q. v.), pp. 57-77.

Cailleux, A., 1968. *Anatomy of the Earth*. McGraw-Hill, New York, 255 pp.

Cailteux, J., 1974. Les sulfures du gisement cuprifere stratiforme de Musoshi, Shaba, Zaire. In: P. Bartholome (Editor) (q. v.), pp. 267-276.

Cameron, E. N., 1943. Origin of sulfides in nickel deposits of Mount Prospect, Connecticut. *Geol. Soc. Amer. Bull.*, *54*, 651-686.

Cann, J. R., 1971. Major element variations in ocean-floor basalts. *Phil. Trans. Roy. Soc. Lond.*, *A*, *268*, 495-505.

Cannon, W. H., 1974. The Chandler annual resonance and its possible geophysical significance. *Phys. Earth Planet. Int.*, *9*, 83-90.

Caputo, M., 1967. *The Gravity Field of the Earth*. Academic, New York, 202 pp.

Carroll, B. M., Simmons, G., and Erickson, A. J., 1974. Sea

floor spreading in the southeast Pacific between 36° and $52^{o}S$. Trans. Amer. Geophys. Union, 55, 301.

Casasola, A. G., 1956. Geological reconnaissance of southern Palawan. Philippine Geol., 10(4).

Cassedanne, J. P., 1972. Les gites Bresiliens de plomb et de zinc. Internat. Geol. Cong. Abstr., 24, 116.

Castle, R. O., and Theodore, T. G., 1972. Some genetic implications of the phase composition of a simple New England pegmatite (GS Prof Ppr 800-B). U. S. Geol. Surv., Washington, pp. B105-B117.

Cawthorn, R. G., and Strong, D. F., 1974. The petrogenesis of komatiites and related rocks as evidence for a layered upper mantle. Earth Planet. Sci. Ltr., 23, 369-375.

Cebull, S. E., and Keller, G. R., 1974. Plate tectonics and the Ouachita system in Texas, Oklahoma, and Arkansas: Reply. Geol. Soc. Amer. Bull., 85, 147-148.

Centeno y Garcia, J., 1882. Informe sobre los temblores de tierra occurridos en el mes de julio de 1879, en el districto de Surigao, isla de Mindanao. Bol. Comis. Mapa Geol. de Espana, 9, 215-221.

Cepeda, J. C., 1973. Evidence for multiple Precambrian deformation in the Sangre de Cristo Mountains, New Mexico. Geol. Soc. Amer. Abstr. Prog., 5, 470.

Challinor, J., 1964. A Dictionary of Geology, 2nd ed. Oxford Univ. Press, New York, 289 pp.

Channell, J. E. T., and Tarling, D. H., 1975. Paleomagnetism and the rotation of Italy. Earth. Planet. Sci. Ltr., 25, 177-188.

Chao, E. C. T., 1973. Geologic implications of the Apollo 14 Fra Mauro breccias and comparison with ejecta from the Ries Crater, Germany. J. Res. U. S. Geol. Surv., 1, 1-18.

Chaudhuri, S., and Cullers, R. L., 1973. Rare earths in clay minerals. Geol. Soc. Amer. Abstr. Prog., 5, 350.

Chemla, M., Touray, J. C., Brousse, R., and Causse, C., 1968. Abondance relative des gas inclus dans quelques mineraux magmatiques d'origine profonde. In: L. H. Ahrens (Editor), Origin and Distribution of the Elements. Pergamon, New York, Pp. 971-179.

Chernyshov, N. M., 1972a. New nickel-bearing province of the south of the Russian Platform. Internat. Geol. Cong. Abstr., 24, 117.

Chernyshov, N. M., 1972b. Precambrian intrusive complexes of basic and ultrabasic rocks in the Verohezh crystalline rock mass with special reference to their common metalliferous features. Proc. USSR Acad. Sci., Geol., 4, 35-47.

Chester, A. H., 1896. A Dictionary of the Names of Minerals. Wiley, New York.

Chesworth, W., 1972. Possible plate contact in the Precambrian of eastern Canada. Nature Phys. Sci., 237, 11-12.

Christensen, N. I., 1970. Composition and evolution of oceanic crust. Marine Geology, 8, 139-154.

Churkin, M. J., 1972. Western boundary of the North American Plate in Asia. Geol. Soc. Amer. Bull., 83, 1027-1056.

Churkin, M. J., 1973. Paleozoic and Precambrian rocks of Alaska and their role in its structural evolution. U. S. Geol. Surv. Prof. Ppr. 740, 62 pp.

Churkin, M., 1975. Basement rocks of Barrow Arch, Alaska, and Circum-Arctic Paleozoic mobile belt. Amer. Assoc. Petrol. Geol. Bull., 59, 451-456.

Ciric, B., 1972. Correlation des phases tectonogenetiques et magmatiques alpines dans les Dinarides. Bull. Soc. Geol. France, Ser. 7, 12, 945-947.

Clague, D. A., and Jarrard, R. D., 1973. Tertiary Pacific plate motion deduced from the Hawaiian-Emperor chain. Geol. Soc. Amer. Bull., 84, 1135-1154.

Clark, K. F., 1972. Stockwork molybednum deposits in the

western Cordillera of North America. Econ. Geol., 67, 731-758.

Clark, A. H., and Farrar, E., 1973. The Bolivian tin province: Notes on the available geochronological data. Econ. Geol., 68, 102-106.

Clark, B. R., and Kelly, W. C., 1973. Sulfide deformation studies: Experimental deformation of pyrrhotite and sphalerite to 2000 bars and 450°C. Econ. Geol., 68, 332-352.

Clark, S. P., Turekian, K. K., and Grossman, L., 1972. Models for the early history of the Earth. In: E. C. Robertson (Editor) (q. v.), pp. 3-18.

Clemmey, H., 1974. Sedimentary geology of a late Precambrian copper deposit at Kitwe, Zambia. In: P. Bartholome (Editor) (q. v.), pp. 255-265.

Cloud, P., 1972. A working model for the primitive Earth. Amer. J. Sci., 272, 537-552.

Cloud, P., 1974. Rubey Conference on crustal evolution. Science, 183, 878-881

Coleman, E. H., 1973. Continental drift and the distribution of fossil reptiles. In: D. H. Tarling and S. K. Runcorn (Editors), (q. v.), vol. 1, pp. 395-412.

Colchester, D. M., 1972. A preliminary note on kimberlite occurrences in South Australia. J. Geol. Soc. Austral., 19, 383-386.

Coleman, R. G., 1971. Plate tectonic emplacement of upper mantle material along continental edges. J. Geophys. Res., 76, 1212-1222.

Coleman, P. U. (Editor), 1973. The Western Pacific: Island Arcs, Marginal Seas, Geochemistry. Univ. Western Australia Press, Perth, 675 pp.

Coleman, P. J., 1975. On island arcs. Earth Sci. Rev., 11, 47-80.

Colley, H., and Warden, A. J., 1974. Petrology of the New

Hebrides. Geol. Soc. Amer. Bull., 85, 1635-1646.

Colony, W. E., and Nordlie, B. E., 1973. Liquid sulfur at Volcan Azufre, Galapagos Islands. Econ. Geol., 68, 371-380.

Comer, J. B., 1973. Genesis of Jamaican bauxite. Geol. Soc. Amer. Abstr. Prog., 5, 251-252.

Coney, P. J., 1972. Cordilleran tectonics and North American plate motion. Amer. J. Sci., 272, 603-628.

Coney, P. J., 1973a. Plate tectonics and marginal foreland thrust-fold belts. Geology, 1, 131-134.

Coney, P. J., 1973b. Non-collision tectonogenesis in western North America. In: D. H. Tarling and S. K. Runcorn (Editors) (q. v.), vol. 2, pp. 713-727.

Cook, P. J., 1972. Petrology and geochemistry of northwest Queensland. Econ. Geol., 67, 1193-1213.

Cook, A. H., and Gaskell, T. F. (Editors), 1961. The Earth Today. Roy. Astron. Soc., London, 404 pp.

Corliss, J. B., 1973. Volcanism and ore-formation related to the subduction of oceanic crust: The "geostill" concept. Geol. Soc. Amer. Abstr. Prog., 5, 26.

Cormier, V. F., 1957. Tectonics near the junction of the Aleutian and Kuril-Kamchatka arcs and a mechanism for middle Tertiary magmatism in the Kamchatka basin. Geol. Soc. Amer. Bull., 86, 443-453.

Coulson, C., and Dupuy, C., 1975. Evolution spatiale des caracteres chimiques du volcanisme andesitique de la Sardaigne (Italie). Earth Planet. Sci. Ltr., 25, 170-176.

Coward, M. P., 1975. Archean deformation patterns in southern Africa. Mtg. Roy. Soc. Lond., Mar. 20-21.

Cox, A., Doell., R. R., and Dalrymple, G. B., 1964. Reversals of the Earth's magnetic field. Science, 144, 1537-1543.

Cox, D. P., Schmidt, R. G., Vine, J. D., Kirkemo, H., Tourtelot, E. B., and Fleischer, M., 1973. Copper. In: D. A. Brobst and W. P. Pratt (Editors) (q. v.), pp. 163-190.

Crawford, A. R., and Wilson, J. T., 1969. Continental drift in the Arctic and Indian oceans and before the Mesozoic era. Geol. Soc. Amer. Abstr. Prog., 7, 41.

Creasey, S. C., 1966. Hydrothermal alteration. In: Titley and Hicks (Editors) (q. v.), pp. 51-74.

Creer, K. M., 1973. A discussion of the arrangement of paleomagnetic poles on the map of Pangaea for epochs in the Phanerozoic. In: D. H. Tarling and S. K. Runcorn (Editors) (q. v.), vol. 1, pp. 47-76.

Cronk, C., 1974. A suggested plate tectonic history of the Indonesia area. Trans. Amer. Geophys. Union, 55, 301.

Crosson, R. S., and Frank, D., 1975. The Mt. Rainier earthquake of July 18, 1973, and its tectonic significance. Bull. Seis. Soc. Amer., 65, 393-401.

Currie, K. L., 1969. Geological notes on the Carswell circular structure, Saskatchewan. Can. Geol. Surv. Ppr. 67-32, 6 pp.

Czamanskee, G. K., 1974. The FeS content of sphalerite along the chalcopyrite-pyrite-bornite sulfur fugacity buffer. Econ. Geol., 69, 1328-1334.

Dagger, G. W., 1972. Genesis of the Mount Pleasant tungsten-molybdenum-bismuth deposit, New Brunswick, Canada. Trans. Inst. Min. Metall., 81, B73-76.

Dallmeyer, R. D., 1975. 40-Ar/39-Ar incremental release ages of biotite from the Cherokee orebody, Ducktown, Tnnnessee: Their bearing on the age of sulfide mineralization. Econ. Geol., 70, 341-345.

Dalrymple, G. B., Silver, E. A., and Jackson, E. D., 1973. Origin of the Hawaiian Islands. Amer. Sci., 61, 294-308.

Dalziel, I. W. D., and Elliott, D. H., 1971. Evolution of Scotia arc. Nature, 233 (5317), 246-252.

Dalziel, I. W. D., Klingfield, R., Lowrie, W., and Opdyke,

N. D., 1973. Paleomagnetic data from the southernmost Andes and the Antarctandes. In: Tarling and Runcorn (Editors) (q. v.), vol. 1, pp. 87-101.

Dana, J. D., 1873. On some results of the Earth's contraction from cooling, including a discussion of the origin of mountains and the nature of the Earth's interior. Amer. J. Sci., Ser. 3, 5, 423-443; 6, 6-14, 104-115, 161-172.

Dandurand, J. L., Fortune, J. P., Perami, R., Schott, J., and Tollon, F., 1972. On the importance of mechanical action and thermal gradient in the formation of metal-bearing deposits. Min. Dep., 7, 339-350.

Dar, K. K., 1968. Metallogeny in the Himalayas. In: M. Stemprok (Editor), Endogenous Ore Deposits. Academia, Prague, pp. 35-42.

Dar, K. K., 1972. Evolution of the Aravelli-Dharwar geosynclinal orogenic belt in India and the development of its structural pattern. Internat. Geol. Cong. Abstr., 24, 72.

Darracott, B. W., 1974. On the crustal structure and evolution of southeastern Africa and the adjacent Indian Ocean. Earth. Planet Sci. Ltr., 24, 282-290.

Dasch, E. J., 1975. Metallogenesis in the southeastern Pacific: A progress report on the IDOE Nazca plate project. Phys. Earth Planet. Int., 9, 249-258.

Davis, G. A., 1973. Subduction-obduction models for the Antler and Sonoma orogenies, western Great Basin area. Geol. Soc. Amer. Abstr. Prog., 5, 592-593.

Davis, J. D., and Guilbert, J. M., 1973. Distribution of the radioelements potassium, uranium, and thorium in selected porphyry copper deposits. Econ. Geol., 68, 145-160.

Davletov, I. K., and Dzhahshibayev, S., 1971. The mineral balance and the behavior of gold during the emplacement of an intrusive body. Geochem. Internat., 1970 (11-12), 997-1011.

Dawes, P. R., 1973. The north Greenland fold belt: A clue to the history of the Arctic Ocean basin and the Nares Strait lineament. In: Tarling and Runcorn (Editors) (q. v.), vol 2, pp. 925-947.

Dean, P. L., and Strong, D. F., 1975. The volcanic stratigraphy, geochemistry, and metallogeny of the central Newfoundland Appalachians. Geol. Soc. Amer. Abstr. Prog., 7, 745-746.

De Bremaecker, J. Cl., 1974. Temperature in a convecting upper mantle. Tectonophys., 21, 1-13.

Degens, E. T., and Ross, D. A. (Editors), 1969. Hot Brines and Recent Heavy Metals in the Red Sea. Springer-Verlag, New York, 600 pp.

De Geoffroy, J., and Wignall, T. K., 1972. A statistical study of geological characteristics of porphyry-copper-molybdenum deposits in the Cordilleran Belt. Econ. Geol., 67, 656-668.

De Jong, K. A., and Scholten, R. (Editors), 1973. Gravity and Tectonics. Wiley, New York, 502 pp.

Delaney, P., and Smith, C. H. (Editors), 1969. Deep-Seated Foundations of Geological Phenomena. Tectonophys., 7, 359-610.

De Launay, L., 1913. Traite de metallogenie: Gites mineraux et metalliferes, vol. 1. Paris.

De Sitter, L. U., 1956. Structural Geology. McGraw-Hill, London, 552 pp.

Deutsch, E. R., and Murthy, G. S., 1973. Late Paleozoic and Triassic paleomagnetism and the opening of the North Atlantic. In: Tarling and Runcorn (Editors) (q. v.), vol. 1, p. 131.

Deutsch, E. R., and Rao, K. V., 1973. Preliminary magnetic study of Newfoundland ophiolites, indicating the presence of native iron. In: Tarling and Runcorn (Editors) (q. v.), vol. 2, pp. 987-992.

Devine, S. B., Ferrell, R. E., and Billings, G. K., 1973. Mineral distribution patterns, deep Gulf of Mexico. Amer. Assoc.

Petrol. Geol. Bull., 57, 28-41.

De Vos, W., Viaene, W., and Moreau, J., 1974. Mineralogie du gisement de Kipushi, Shaba, Zaire. In: Bartholome (Editor) (q. v.), pp. 165-183.

Dewey, J. F., and Bird, J. M., 1970a. Mountain belts and the new global tectonics. J. Geophys. Res., 75, 2625-2647.

Dewey, J. F., and Bird, J. M., 1970b. Plate tectonics and geosynclines. Tectonophys., 10, 6250638.

Dewey, J. F., and Burke, K., 1973. Plume generated triple junctions. Trans. Amer. Geophys. Union, 54, 229.

Dewey, J. F., Pittman, W. C., III, Ryan, W. F. B., and Bonnin, J., 1973. Plate tectonics and the evolution of the Alpine system. Geol. Soc. Amer. Bull., 84, 3137-3180.

Dickinson, W. R., 1972. Evidence for plate-tectonic regimes in the rock record. Amer. J. Sci., 272, 551-576.

Dickinson, W. R., 1973a. Widths of modern arc-trench gaps proportional to past duration of igneous activity in associated magmatic arcs. J. Geophys. Res., 78, 3376-3389.

Dickinson, W. R., 1973b. Reconstruction of past arc-trench systems from petrotectonic assemblages in the island arcs of the western Pacific. In: Coleman (Editor) (q. v.), pp. 569-601

Dietz, R. S., 1952. Geomorphic evolution of continental terrace (continental shelf and slope). Bull. Amer. Assoc. Petrol. Geol., 36, 1802-1819.

Dietz, R. S., 1963a. Wave base, marine profile of equilibrium, and wave-built terraces: A critical appraisal. Geol. Soc. Amer. Bull., 74, 971-990.

Dietz, R., 1963b. Collapsing continental rises: An actualistic concept of geosynclines and mountain building. J. Geol., 71, 314-333.

Dietz, R. S., and Holden, J. C., 1966. Miogeosynclines in space and time. J. Geol., 74, 566-583.

Dietz, R. S., and Holden, J. C., 1970. East Indian basin (Wharton basin) as pre-Mesozoic ocean crust. Geol. Soc. Amer. Abstr. Prog., 2, 537.

Dietz, R. S., and Holden, J. C., 1973. Continents adrift: New orthodoxy of persuasive joker? In: Tarling and Runcorn (Editors) (q. v.), vol. 2, pp. 1105-1121.

Dietz, R. S., and Menard, W. H., 1951. Origin of abrupt change in slope at continental shelf margin. Amer. Assoc. Petrol. Geol. Bull., 35, 1994-2016.

Dietz, R. S., Holden, J. C., and Sproll, W. P., 1969. Geotectonic evolution and subsidence of Bahama Platform. Geol. Soc. Amer. Abstr. Prog., 7, 48.

Dimanche, F., 1974. Paragenese des sulfures de cuivre dans les gisements du Shaba (Zaire): I, Kipushi; II, Kamoto. In: Bartholome (Editor) (q. v.), pp. 185-201.

Dimroth, E., Baragar, W. R. A., Bergeron, R., and Jackson, G. D., 1970. The filling of the Circum-Ungava geosyncline. Geol. Surv. Can. Ppr. 70-40, 45-142.

Dixon, C. J., and Pereira, J., 1973. Plate tectonics and and mineralization in the Middle East. Econ. Geol., 68, 137.

Dixon, C. J., and Pereira, J., 1974. Plate tectonics and mineralization in the Tethyan region. Min. Dep., 9, 185-198.

Dmitriev, L. V., Barsukov, V. L., and Durasova, N. A., 1970. On the question of source of ore substance. In: Pouba and Stemprok (Editors) (q. v.), pp. 8-11.

Dodge, F. C. W., 1972. Trace element contents of some plutonic rocks of the Sierra Nevada batholith (Bull. 1314-F). U. S. Geol. Surv., Washington, 13 pp.

Doe, B. R., and Stacey, J. S., 1974. The application of lead isotopes to the problems of ore genesis and ore prospect evaluation: A review. Econ. Geol., 69, 757-776.

Doell, R. R., and Cox, A., 1972. The Pacific geomagnetic

secular anomaly and the question of lateral uniformity in the lower mantle. In: Robertson (Editor) (q. v.), pp. 248-284.

Donaldson, J. A., McGlynn, J. C., Irving, E., and Park, J. K., 1973. Drift of the Canadian Shield. In: Tarling and Runcorn (Editors) (q. v.), vol. 1, pp. 3-17.

Dott, R. H., Jr. (Editor), 1972. Modern and Ancient Geosynclinal Sedimentation. Univ. Wisc., Madison, 59 pp.

Dott, R. H., Jr., and Shaver, R. H. (Editors), 1974. Modern and Ancient Geosynclinal Sedimentation. Amer. Assoc. Petrol. Geol., Tulsa, 380 pp.

Douglas, R. J. W., 1969. Orogeny, basement, and geological map of Canada. Geol. Assoc. Can. Spl. Ppr. 5, 1-6.

Douglas, R. J. W., 1970. Geology and Economic Minerals of Canada. Geol. Surv. Can., Ottawa, 838 pp.

Douglas, R. G., Moullade, M., and Nairn, A. E. M., 1973. Cause and consequences of drift in the South Atlantic. In: Tarling and Runcorn (Editors) (q. v.), vol. 1, pp. 517-537.

Drake, C. L., 1969. Continental margins. In: Hart (Editor) (q. v.), pp. 549-556.

Drake, C. L., 1970. A long-range program of solid Earth studies. Trans. Amer. Geophys. Union, 51, 152-159.

Drake, C. L. (Editor), 1973. U. S. Program for the Geodynamics Project: Scope and Objectives. Nat. Acad. Sci., Wash., 235 pp.

Drake, C. L., and Kosminskaya, I. P., 1969. The transition from continental to oceanic crust. Tectonophys., 7, 363-384.

Drake, C. L., Ewing, M., and Sutton, G. H., 1959. Continental margins and geosynclines; the east coast of North America north of Cape Hatteras. In: Ahrens et al. (Editors) (q. v.), vol. 3, pp. 110-198.

Drake, C. L., Ewing, J. L., and Stockland, H., 1968. The continental margin of the eastern United States. Can. J. Earth

Sci., *5*, 993-1010.

Dubois, J., Launay, J., and Recy, J., 1975. Some new evidence on lithospheric bulges close to island arcs. *Tectonophys.*, *26*, 189-196.

Dunbar, C. O., 1966. *The Earth*. World, New York, 252 pp.

Duncan, R. A., and McDougall, I., 1974. Migration of volcanism with time in the Marquesas Islands, French Polynesia. *Earth Planet. Sci. Ltr.*, *21*, 414-420.

Dunnett, D., 1975. Precambrian deformation patterns in Australia. Mtg. Roy. Soc. Lond., Mar. 20-21.

Durand, B. J., and Espitalie, J., 1973. Evolution de la materie organique au cours de l'enfouissement des sediments. *Acad. Sci. Comp. Rend.*, *276*, 2253-2256.

Durkee, E. F., and Pedersen, S. L., 1961. Geology of southern Luzon, Philippines. *Amer. Assoc. Petrol. Geol. Bull.*, *45*, 137-168.

Dymond, J., 1975. K-Ar ages of Tahiti and Moorea, Society Islands, and implications for the hot-spot model. *Geology*, *3*, 236-240.

Dzotsenidze, G. S., 1972. On the volcanic-sedimentary iron ore deposits. *Internat. Geol. Cong. Abstr.*, *24*, 119-120.

Eardley, A. J., 1963. Relation of uplifts to thrusts in Rocky Mountains. In: *Backbone of the Americas: Tectonic History from Pole to Pole.* Amer. Assoc. Petrol. Geol., Mem. 2.

Easton, W. H., and Melandres, M. M., 1963. First Paleozoic fo il from Philippine archipelago. *Amer. Assoc. Petrol. Geol. Bull.*, *47*, 1871-1886.

Elsass, P., 1974. Structures sedimentaires et mineralisations dans le gite de Talate n'Ouamane Anti-Atlas occidental, Maroc. In: Bartholome (Editor) (q. v.), pp. 331-338.

Elsasser, W. M., 1966. Thermal structure of the upper mantle. In: P. M. Hurley (Editor), *Advances in Earth Sciences*. MIT Press,

Cambridge, pp. 461-471.

Elsasser, W. M., 1967. Interpretation of heat flow equality. J. Geophys. Res., 72, 4768-4770.

Elston, D. P., and Gromme, C. S., 1974. Precambrian Polar wandering from Unkar Group and Nankoweap Formation, eastern Grand Canyon, Arizona. Geol. Soc. Amer. Abstr. Prog., 6, 440-441.

Emery, K. O., 1965. Characteristics of continental shelves and slopes. Amer. Assoc. Petrol. Geol. Bull., 49, 1379-1384.

Emery, K. O., 1968. Relict sediments on continental shelves of the world. Amer. Assoc. Petrol. Geol. Bull., 52, 445-464.

Emery, K. O., and Uchupi, E., 1972. Western North Atlantic Ocean. Amer. Assoc. Petrol. Geol., Tulsa, 532 pp.

Engels, A. E. J., and Kelm, D. L., 1972. Pre-Permian global tectonics: A tectonic test. Geol. Soc. Amer. Bull., 83, 2325-2340.

England, R. N., and Davis, H. L., 1973. Mineralogy and ultramafic cumulates and tectonites from eastern Papua. Earth Planet. Sci. Ltr., 17, 416-425.

Ernst, W. G., 1972. Occurrence and mineralogic evolution of blueschist belts with time. Amer. J. Sci., 272, 657.

Ernst, W. G., 1973a. Blueschist metamorphism and P-T regimes in active subduction zones. Tectonophys., 17, 255-272.

Ernst, W. G., 1973b. Interpretive synthesis of metamorphism in the Alps. Geol. Soc. Amer. Bull., 84, 2053-2078.

Ernst, W. G., 1975. Systematics of large-scale tectonics and age progressions in Alpine and Circum-Pacific blueschist belts. Tectonophys., 26, 229-246.

Escher, A., Jack, S., and Watterson, J., 1975. Brittle deformation, dyke intrusion and the mechanical characteristics of Early Proterozoic crust in the North Atlantic region. Mtg. Roy. Soc. Lond., Mar. 13-14.

Ewers, W. E., and Hudson, D. R., 1972. An interpretative study of a nickel-iron sulfide ore intersection, Lunnon Shoot,

Kambalda, Western Australia. Econ. Geol., 67, 1075-1092.

Ewing, J., and Ewing, M., 1967. Sediment distribution on the mid-ocean ridges with respect to the spreading sea floor. Science, 156, 1590-1592.

Fahrig, W. F., Irving, E., and Jackson, J. D., 1971. Paleomagnetism of the Franklin diabases. Can. J. Earth Sci., 8, 455-467.

Fairbridge, R. W., 1973. Glaciation and plate migration. In: Tarling and Runcorn (Editors) (q. v.), vol. 1, pp. 503-515.

Fardon, R. S. H., 1971. The Western Australian nickel deposits. In: J. D. Glover (Editor), Symposium on Archean Rocks. Geol. Soc. Austral. Spl. Publ. 3, p. 256.

Farquharson, R. B., and Richards, J. R., 1974. U-Th-Pb isotope systematics related to igneous rocks and ore Pb, Mount Isa, Queensland. Min. Dep., 9, 339-356.

Faure, H., 1972. Paleodynamique du craton africain. Internat. Geol. Cong. Abstr., 24, 75.

Faure, H., 1973. Vertical movements and horizontal translations of the lithosphere. In: Tarling and Runcorn (Editors) (q. v.), vol. 2, pp. 731-733.

Faure, G., and Powell, J. L., 1972. Strontium Isotope Geology. Springer, New York, 188 pp.

Fein, C. D., 1973. Chemical composition and submarine weathering of some oceanic tholeiites from the East Pacific Rise. Geol. Soc. Amer. Abstr. Prog., 5, 41.

Ferguson, J., and Lambert, I. B., 1972. Volcanic exhalations and metal enrichments at Matupi Harbor, New Britain, T. P. N. G., Econ. Geol., 67, 25-37.

Field, C. W., Jones, M. B., and Bruce, W. R., 1973. Porphyry copper-molybdenum deposits of the Pacific northwest. Econ. Geol., 68, 137.

Fischer, A. G., 1973. Sedimental logic and stratigraphic patterns. Trans. Amer. Geophys. Union, 54, 329.

Fisher, W. L., Proctor, C. V., Galloway, W. E., and Nagle, J. S., 1970. Depositional system in the Jackson Group of Texas, their relationship to oil, gas, and uranium. Gulf Coast Assoc. Geol. Soc. Trans., 20, 234-261.

Fitch, F. H., 1958. The geology and mineral resources of the Sandakan area, North Borneo. Geol. Surv. Dept. Brit. Terr. Borneo Mem. 9, 1-202.

Fitch, F. H., 1963. Geological relationship between the Philippines and Borneo. Philippine Geol., 17, 41-47.

Fitch, T. J., and Scholz, C. H., 1971. Mechanism of underthrusting in southwest Japan: A model of convergent plate interactions. J. Geophys. Res., 76, 7260-7292.

Fleischer, R., Routhier, P., 1973. The "consanguineous" origin of a tourmaline-bearing gold deposit: Passagem de Mariana (Brazil). Econ. Geol., 68, 11-22.

Flemming, N. C., and Roberts, D. G., 1973. Tectoniceustatic changes in sea level and sea floor spreading. Nature, 243, 19-22.

Flinter, B. H., Hesp, W. R., and Rigby, D., 1972. Selected geochemical, mineralogical, and petrological features of granitoids of the New England complex, Australia, and their relation to Sn, W, Mo, and Cu mineralization. Econ. Geol., 67, 1241-1262.

Forsyth, D. W., 1975. Fault plane solutions and tectonics of the South Atlantic and Scotia Sea. J. Geophys. Res., 80, 1429-1443.

Fountain, R. J., 1972. Geological relationships in the Panguna porphyry copper deposit, Bougainville Island, New Guinea. Econ. Geol., 67, 1049-1064.

Francheteau, J., 1973. Plate tectonic model of the opening of the Atlantic Ocean south of the Azores. In: Tarling and Runcorn (Editors) (q. v.), vol. 1, pp. 197-202.

Francheteau, J., and Le Pichon, X., 1972. Marginal fracture zones as structural framework of continental margins in South Atlantic Ocean. Amer. Assoc. Petrol. Geol. Bull., 56, 991-1007.

Francheteau, J., Harrison, C. G. A., Sclater, J., and Richards, M. L., 1969. Magnetization of Pacific seamounts: a preliminary polar curve for the northeastern Pacific. Trans. Amer. Geophys. Union, 50, 634.

Francois, A., 1974. Stratigraphie, tectonique et mineralisations dans l'arc cuprifere du Shaba (Republique du Zaire). In: Bartholome (Editor) (q. v.), pp. 79-101.

Franklin, J. M., Kasarda, J., and Poulsen, K. H., 1975. Petrology and chemistry of the alteration zone of the mattabi massive sulfide deposit. Econ. Geol., 70, 63-79.

Frazer, M. C., 1974. Spherical harmonic analysis of the Navier-Stokes equation in magnetofluid dynamics. Phys. Earth Planet. Inter., 8, 75-82.

Freund, R., 1974. Kinematics of transform and transcurrent faults. Tectonophys., 21, 93-134.

Gaal, G., 1972. Tectonic control of some Ni-Cr deposits in Finland. Internat. Geol. Cong. Abstr., 24, 122.

Gabelman, J. W., and Krusiewski, S. V., 1972. The metallotectonics of Europe. Internat. Geol. Cong. Abstr., 24, 122.

Gabrielse, H., 1972. Younger Precambrian of the Canadian Cordillera. Amer. J. Sci., 272, 521-536.

Gandhi, S. S., and Brown, S. S., 1975. Cupriferous shales of the Adeline Island formation, Seal Lake group, Labrador. Econ. Geol., 70, 145-163.

Garland, G. D. (Editor), 1966. Continental Drift. Univ. Toronto Press, Toronto, 140 pp.

Garrels, R. M., Mackenzie, F. T., and Silver, R., 1972. Sedimentary cycling in relation to the history of the continents and oceans. In: Robertson (Editor) (q. v.), pp. 93-121.

Gaskell, T. F. (Editor), 1967. The Earth's Mantle. Academic, London, 509 pp.

Gasparrini, E., and Naldrett, A. J., 1972. Magnetite and

ilmenite in the Sudbury nickel irruptive. Econ. Geol., 67, 605-621.

Gast, P. W., 1972. The chemical composition of the Earth, the Moon, and chondritic meteorites. In: Robertson (Editor) (q. v.), pp. 19-40.

Gastil, G., 1960. The distribution of mineral dates in time and space. Amer. J. Sci., 258, 1.

Gastil, R. G., and Krummenacher, D., 1975. Reconstruction of the Mesozoic plutonic belt of southern California and western Mexico. Geol. Soc. Amer. Abstr. Prog., 7, 320.

Gealy, E. L., 1971. Results of the JOIDES Deep Sea Drilling Project: 1968-1971. Eighth World Petroleum Congress Proceedings, vol. 2. Applied Science Publishers, London, pp. 337-348.

Gehlen, K. V., and Forkel, W., 1972. Chromite-bearing complexes, anomalous mantle, and plate tectonics. Internat. Geol. Cong. Abstr., 24, 122-123.

Gilbert, F., 1972. Inverse problems for the Earth's normal modes. In: Robertson (Editor) (q. v.), pp. 125-146.

Gill, R. C. O., and Bridgwater, D., 1975. Geochemistry and significance of early Archean low-K and related tholeiite dykes cutting granitic gneisses in West Greenland. Geol. Soc. Amer. Abstr. Prog., 7, 763.

Gilluly, J., 1971. Plate tectonics and magmatic evolution. Geol. Soc. Amer. Bull., 82, 2383-2396.

Gilluly, J., 1972. Tectonics involved in the evolution of mountain ranges. In: Robertson (Editor) (q. v.), pp. 406-439.

Glikson, A. Y., 1970. Geosynclinal evolution and geochemical affinities of early Precambrian systems. Tectonophys., 9, 397-433.

Gnibidenko, H. S., and Shashkin, K. S., 1970. Basic principles of the geosynclinal theory. Tectonophys., 9, 5-13.

Goldreich, P., and Toomre, A., 1969. Some remarks on polar wandering. J. Geophys. Res., 74, 2555-2567.

Goldschmidt, V. M., 1954. Geochemistry. Clarendon, Oxford, 730 pp.

Goodwin, A. M., 1968. Archean protocontinental growth and early crustal history of the Canadian shield. In: Benes (Editor) (q. v.), pp. 69-89.

Goodwin, A. M., 1973. Plate tectonics and evolution of Precambrian crust. In: Tarling and Runcorn (Editors) (q. v.), vol. 2, pp. 1047-1069.

Goodwin, A. M., 1975. Archean volcanism in Superior Province, Canadian Shield. Geol. Soc. Amer. Abstr. Prog., 7, 764.

Goossens, P. J., 1972. Metallogeny in Ecuadorian Andes. Econ. Geol., 67, 458-468.

Goossens, P. J., 1973. On the porphyry copper age determinations: a reply. Econ. Geol., 68, 284.

Gosh, A. K., 1972. Trace element geochemistry and genesis of the copper ore deposits of the Singhbhum shear zone, eastern India. Min. Dep., 7, 292-313.

Gough, D. I., 1970. Did an ice cap break Gondwanaland? J. Geophys. Res., 75, 4475-4477.

Graham, S. A., Dickinson, W. R., and Ingersoll, R. V., 1975. Himalayan-Bengal model for flysch dispersal in the Appalachian-Ouachita system. Geol. Soc. Amer. Bull., 86, 273-286.

Graham, S. A., Dickinson, W. R., and Ingersoll, R. V., 1974. Himalayan-Bengal vs. Appalachian-Ouachita analogy and patterns of synorogenic sedimentation, in press.

Granath, J. W., 1975. Wind River Canyon: An example of a greenstone belt in the Archean of Wyoming, U. S. A. Precamb. Res., 2, 71-91.

Green, D. H., 1969. The origin of basaltic and nephelinitic magmas in the Earth's mantle. Tectonophys., 7, 409-422.

Green, D. H., 1975. Genesis of Archean peridotitic magmas and constraints of Archean geothermal gradients and tectonics.

Geology, 3, 15-18.

Green, D. H., Nicholls, I. A., Viljoen, M., and Viljoen, R., 1975. Experimental demonstration of the existence of peridotitic liquids in earliest Archean magmatism. Geology, 3, 11-14.

Greenwood, W. R., Hadley, D. G., Anderson, R. E., Fleck, R. J., and Schmidt, D. L., 1975. Late Proterozoic cratonization in the southern part of the Arabian shield. Mtg. Roy. Soc. Lond., Mar. 13-14.

Griffin, V. S., Jr., 1970. Relevancy of the Dewey-Bird hypothesis of cordilleran-type mountain belts and the Wegmann stockwork concept. J. Geophy. Res., 75, 7504-7507.

Griffin, V. S., Jr., 1974. Plate tectonics and the Ouachita system of Texas, Oklahoma, and Arkansas: Discussion. Geol. Soc. Amer. Bull., 85, 145-146.

Griffis, R., 1972. Genesis of magnesite deposits, Deloro Twp., Ontario. Econ. Geol., 67, 63-71

Griffitts, W. R., Albers, J. P., and Oner, O., 1972. Massive sulfide copper deposits of the Ergani-Madan area southeastern Turkey. Econ. Geol., 67, 701-716.

Griggs, D. T., 1939. A theory of mountain building. Amer. J. Sci., 237, 611-650.

Griggs, D. T., 1972. The sinking lithosphere and the focal mechanism of deep earthquakes. In: Robertson (Editor) (q. v.), pp. 361-384.

Gromme, C. S., Merrill, R. T., and Verhoogen, J., 1967. Paleomagnetism of Jurassic and Cretaceous plutonic rocks in the Sierra Nevada, California, and its significance for polar wandering and continental drift. J. Geophys. Res., 72, 5661-5684.

Groves, D. I., and Baker, W. E., 1972. The regional variation in composition of wolframites from Tasmania. Econ. Geol., 67, 362-368.

Groves, D. I., Hudson, D. R., and Hack, T. B. C., 1969.

Modification of iron-nickel sulfides during serpentinization and talc-carbonate alteration at Black Swan, Australia. Econ. Geol., 69, 1265-1281.

Grow, J. A., 1973. Crustal and upper mantle structure of the central Aleutians arc. Geol. Soc. Amer. Bull., 84, 2169-2192.

Grow, J. A., and Bowin, C. O., 1975. Evidence for high-density crust and mantle beneath the Chile trench due to the descending lithosphere. J. Geophys. Res., 80, 1449-1458.

Guilbert, J. M., 1972. Mineral deposit exploration, new global tectonics, and new thinking. Internat. Geol. Cong. Abstr., 24, 125.

Guild, P. W., 1972. Metallogeny and the new global tectonics. Internat. Geol. Cong. Abstr., 24, 126.

Guild, P. W., 1973. Massive sulfide deposits as indicators of former plate boundaries. Econ. Geol., 68, 137-138.

Guilloux, L., and Pelissonnier, H., 1974. Les gisements de schistes, marnes et gres cupriferes. In: Bartholome (Editor) (q. v.), pp. 35-55.

Gulson, B. L., and Krogh, T. E., 1975. Evidence of multiple intrusion, possible resetting of U-Pb ages, and new crystallization of zircons in the post-tectonic intrusions ("Rapakivi granites") and gneisses from South Greenland. Geochim. Cosmochim. Acta, 39, 65-82.

Gurbanov, A. G., 1972. New data on the geological patterns and age of the Kardyvach intrusion in the Caucasus. Proc. USSR Acad. Sci. Geol., 9, 24-39.

Gutenberg, B., 1951. Internal Constitution of the Earth. Dover, New York, 439 pp.

Gutenberg, B., and Richter, C. F., 1954. Seismicity of the Earth, 2nd ed. Princeton Univ. Press, Princeton.

Hackett, J. P., and Bischoff, J. L., 1973. New data on the stratigraphy, extent, and geologic history of the Red Sea geothermal deposits. Econ. Geol., 68, 553-564.

Haile, N. S., 1973. The recognition of former subduction zones in southeast Asia. In: Tarling and Runcorn (Editors) (q. v.), vol. 2, pp. 885-892.

Haile, N. S., and Wong, N. P. Y., 1965. The geology and mineral resources of Dent peninsula, Sabah. Geol. Surv. Borneo Reg. Malaysia, Mem. 16, 1-199.

Hailwood, E. A., and Tarling, D. H., 1973. Paleomagnetic evidence for a proto-Atlantic Ocean. In: Tatling and Runcorn (Editors) (q. v.), vol. 1, pp. 37-46.

Hales, A. L., 1974. Crustal structure modelling in a spherically symmetrical Earth. Phys. Earth Planet. Int., 9, 7-8.

Hales, A. L., and Herrin, E., 1972. Travel times of seismic waves. In: Robertson (Editor) (q. v.), pp. 172-215.

Hall, J., 1857. Direction of currents of deposition and sources of the materials of the older Paleozoic rocks. Can. Naturalist and Geologist, 2, 284-286.

Hallam, A., 1971. Mesozoic geology and the opening of the North Atlantic. J. Geol., 79, 129-157.

Hallam, A., 1973. Provinciality, diversity, and extinction of Mesozoic marine invertebrates in relation to plate movements. In: Tarling and Runcorn (Editors) (q. v.), vol. 1, pp. 287-294.

Hallberg, J. A., 1972. Geochemistry of Archean volcanic belts in the Eastern Goldfields region of Western Australia. J. Petrol., 13, 45-56.

Hamilton, W., 1973. Tectonics of the Indonesian region. In: Tarling and Runcorn (Editors) (q. v.), vol. 2, pp. 879-884.

Hammond, S. R., Kroende, L. W., and Thayer, F., 1974. Latitudinal displacements on the Ontong Java Plateau between 30 and 100 m. y. ago. Geol. Soc. Amer. Abstr. Prog., 6, 374-5.

Handschumacher, D. W., and Andrews, J. E., 1975. Kana Keoki fracture zone: interaction with the Hawaiian Ridge. Geology, 3, 25-28.

Hanor, J. S., 1973. Regional control of Mississippi Valley type ore districts in eastern North America. Geol. Soc. Amer. Abstr. Prog., 5, 320.

Harrison, J. E., 1972. Precambrian Belt basin of northwestern United States: its geometry, sedimentation, and copper occurrences. Geol. Soc. Amer. Bull., 83, 1215-1240.

Harrison, J. E., 1974. Copper mineralization in miogeosynclinal clastics of the Belt Supergroup, northwestern United States. In: Bartholome (Editor) (q. v.), pp. 353-366.

Harrison, C. G. A., Bonatii, E., Honnorez, J., Fisher, D. E., and Salawan, J., 1974. The Easter Island seamount chain and the non-hot spot hypothesis. Trans. Amer. Geophys. Union, 55, 300.

Hart, P. J. (Editor), 1969. The Earth's Crust and Upper Mantle: Structure, Dynamic Processes, and Their Relation to Deep-Seated Geological Processes. Amer. Geophys. Union, Washington, 735 pp.

Haskin, L. A., Haskin, M. A., Frey, F. A., and Wildeman, T. R., 1968. Relative and absolute terrestrial abundances of the rare earths. In: Ahrens (Editor) (q. v.), pp. 889-912.

Hawkesworth, C. J., Waters, D. J., and Bickle, M. J., 1975a. Plate tectonics in the eastern Alps. Earth Planet. Sci. Ltr., 24, 405-413.

Hawkesworth, C. J., Moorbath, S., O'Nions, R. K., and Wilson, J. F., 1975b. Age relationships between greenstone belts and "granites" in the Rhodesian Archean craton. Earth Planet. Sci. Ltr., 25, 251-262.

Hawkins, J. W., 1973. Petrology of marginal basins and their possible significance in orogenic belts: The Lau Basin as an example. Geol. Soc. Amer. Abstr. Prog., 5, 51.

Hawkins, J. W., and Natland, J. H., 1975. Nephelinites and basanites of the Somoan linear volcanic chains: Their possible tectonic significance. Earth Planet. Sci. Ltr., 24, 427-439.

Hayatsu, A., and Palmer, H. C., 1975. K-Ar isochron study of the Tudor Gabbro, Grenville Province, Ontario. Earth. Planet. Sci. Ltr., 25, 208-212.

Haynes, S. J., 1974. Evolution of the Zagros suture zone, southern Iran. Geol. Soc. Amer. Bull., 85, 739-744.

Heacock, J. G. (Editor), 1971. The Structure and Physical Properties of the Earth's Crust. Amer. Geophys. Union, Washington, 384 pp.

Heezen, B. C., and 11 co-authors, 1973. Diachronous deposits, a kinematic interpretation of the post-Jurassic sedimentary sequence on the Pacific plate. Nature, 241, 25-32.

Hein, J. R., 1973. Deep-sea sediment sources areas: Implications of variable rates of movement between California and the Pacific plate. Nature, 241, 62-64.

Heirtzler, J. R. (Editor), 1972. Understanding the Mid-Atlantic Ridge. National Academy of Science, Washington, 131 pp.

Heirtzler J. R., 1973. The evolution of the North Atlantic Ocean. In: Tarling and Runcorn (Editors) (q. v.), vol. 1, pp. 191-196.

Heirtzler, J. R., Dickson, G. O., Herron, E. M., Pitman, W. C., and Le Pichon, X., 1968. Marine magnetic anomalies, geomagnetic field reversals, and motions of the ocean floor and continents. J. Geophys. Res., 73, 2119-2136.

Heiskanen, W. A., and Moritz, H., 1967. Physical Geodesy. Freeman, San Francisco, 364 pp.

Heiskanen, W. A., and Vening Meinesz, F. A., 1958. The Earth and Its Gravity Field. McGraw-Hill, New York, 470 pp.

Helsey, C. E., and Steiner, M. B., 1969. Evidence for long intervals of normal polarity during the Cretaceous period. Earth Planet. Sci. Ltr., 5, 325-332.

Helsey, C. E., and Steiner, M. B., 1974. Paleomagnetism of the Lower Triassic Moenkopi Formation. Geol. Soc. Amer. Bull., 85, 457-464.

Henley, K. J., and Brown, R. N., 1974. Cupriferous hydrobiotite from Ukaparinga, South Australia. Econ. Geol., 69, 688-692.

Herrin, E., 1972. A comparative study of upper mantle models: Canadian Shield and Basin and Range Province. In: Robertson (Editor) (q. v.), pp. 216-231.

Hess, H. H., 1955. Serpentinites, orogeny, and epeirogeny. Geol. Soc. Amer. Spl. Ppr. 62, 341-408.

Hewett, D. F., 1972. Magnanite, hausmannite, braunite: Features, modes of origin. Econ. Geol., 67, 83-102.

Hey, R. N., Deffeyes, K. S., Johnson, G. L., and Lowrie, A., 1972. The Galapagos triple junction and plate motions in the east Pacific. Nature, Phys. Sci., 237, 20-22.

Heyl, A. V., Landis, G. P., and Zartman, R. E., 1974. Isotopic evidence for the origin of Mississippi Valley type mineral deposits: A review. Econ. Geol., 69, 992-1006.

Hills, E. S., 1972. Time and space relationships of basins of epeirogenic origin in Australia. Internat. Geol. Cong. Abstr. 24, 80.

Hills, F. A., Houston, R. S., and Subbarayudu, G. V., 1975. Possible Proterozoic plate boundary in southern Wyoming. Geol. Soc. Amer. Abstr. Prog., 7, 614.

Ho, C. S., 1961. Geologic relationships and comparisons between Taiwan and the Philippines. Philippine Geol., 15(2), 59-95.

Hoagland, A. D., 1973. Appalachian zinc-lead and the deposits of middle Tennessee. Geol. Soc. Amer. Abstr. Prog., 5, 404.

Hodder, R. W., and Hollister, V. F., 1972. Structural features of porphyry copper deposits and the tectonic evolution of continents. Can. Min. Metall. Bull., 65, 41-55.

Hoering, T. C., 1967. The organic geochemistry of Precambrian rocks. In: P. H. Abelson (Editor), Researches in Geochemistry, vol. 2. Wiley, New York, pp. 87-111.

Hoffman, P., 1973. Evolution of an early Proterozoic continental margin: The Coronation geosyncline and associated aulacogens of the northwestern Canadian shield. Roy. Soc. Lond. Phil. Trans., A-273, 547-581.

Hoffman, P. F., 1975. Proterozoic post-orogenic volcanism, Great Bear Lake, NWT. Geol. Soc. Amer. Abstr. Prog., 7, 785-786.

Holland, H. D., 1968. The abundance of CO_2 in the Earth's atmosphere through geologic time. In: Ahrens (Editor) (q. v.), pp. 949-954.

Holland, H. D., 1972. Granites, solutions, and base metal deposits. Econ. Geol., 67, 281-301.

Hollister, V. F., 1973. Regional characteristics of porphyry copper deposits of South America. Econ. Geol., 68, 138.

Hollister, V. F., and Sirvas, E. B., 1974. The Michiquillay porphyry copper deposit. Min. Dep., 9, 261-269.

Hollister, V. F., Potter, R. R., and Barker, A. L., 1974. Porphyry-type deposits of the Appalachian orogen. Econ. Geol., 69, 618-630.

Holmes, A., 1965. Principles of Physical Geology. Ronald, New York, 1288 pp.

Hoover, W. B., 1974. New tectonics theory has origin in convection cells. World Oil, 178(6), 104-106.

Hopson, C. A., and Mattison, J. M., 1973. Ordovician and Late Jurassic ophiolitic assemblages in the Pacific Northwest. Geol. Soc. Amer. Abstr. Prog., 5, 57.

Horn, D. R., Delach, M. N., and Horn, B. M., 1973a. Metal Content of Ferromanganese Deposits of the Oceans. National Sci. Foundation, Washington, 51 pp.

Horn, D. R., Horn, B. M., and Delach, M. N., 1973b. Ocean Manganese Nodules Metal Values and Mining Sites. National Sci. Foundation, Washington, 57 pp.

Houston, M. H., Jr., and De Bremaecker, J. Cl., 1975. Numerical models of convection in the upper mantle. J. Geophys. Res.,

80, 742-751.

Hoyt, D. E., and Scholle, P. A., 1974. Paleogeographic interpretation of the Palmetto Complex, Ordovician of southwest Nevada. Geol. Soc. Amer. Abstr. Prog., 6, 449.

Hsü, K. J., 1965. Collapsing continental rises: a discussion. J. Geophys. Res., 73, 897-900.

Hsü, K. J., 1968. Principles of melanges and their bearing on the Franciscan-Knoxville problem. Geol. Soc. Amer. Bull., 79, 1063.

Hubbard, F. H., 1975. Precambrian crustal development in western Nigeria: Indications from the Iwo region. Geol. Soc. Amer. Bull., 86, 548-554.

Huestis, S., Molnar, P., and Oliver, J., 1973. Regional Sn velocities and shear velocity in the upper mantle. Bull. Seis. Soc. Amer., 63, 469-476.

Hurley, P. M., 1969. Some observations on the geological history of Laurasia. Geol. Soc. Amer. Abstr. Prog., 7, 112.

Hurley, P. M., 1973. On the origin of 450 ± 200 m. y. orogenic belts. In: Tarling and Runcorn (Editors) (q. v.), vol. 2, pp. 1083-1089.

Hurley, P. M., and Rand, J. R., 1969. Pre-drift continental nuclei. Science, 164, 1229-1242.

Hurley, P. M., de Almeida, F. F. M., Melcher, G. C., Cordani, U. G., Rand, J. R., Kawashita, K., Vandoros, P., Pinson, W. H., and Fairbairn, W. H., 1967. Test of continental drift by comparison of radiometric ages. Science, 157, 495-500.

Hutchinson, R. W., and Hodder, R. W., 1972. Possible tectonic and metallogenic relationships between porphyry copper and massive sulphide deposits. Can. Min. Metall. Bull., 65, 34-40.

Intiomale, M. M., and Oosterbosch, R., 1974. Geologie et geochemie du gisement de Kipushi, Zaire. In: Bartholome (Editor) (q. v.), pp. 123-164.

Irving, E. M., 1949. Notes on the geology of Balabac Island, Palawan. Philippine Geol., 3(4), 14-22.

Irving, E. M., 1950. Review of Philippine basement geology and its problems. Philippine J. Sci., 79, 267-307.

Irving, E. M., 1951. Submarine morphology of the Philippine archipelago and its geologic significance. Philippine J. Sci., 80, 55-88.

Irving, E., 1966. Paleomagnetism of some carboniferous rocks from New South Wales and its relation to geological events. J. Geophys. Res., 71, 6025-6051.

Irving, E., and McGlynn, J. C., 1975. Precambrian paleomagnetism of the northern continents and its tectonic implication. Mtg. Roy. Soc. Lond., Mar. 13-14.

Irving, E., Emslie, R. F., and Ueno, H., 1974. Upper Proterozoic paleomagnetic poles from Laurentia and the history of the Grenville structural province. J. Geophys. Res., 79, 5491-5502.

Isaacson, P. E., 1975. Evidence for a western extracontinental land source during the Devonion period in the Central Andes. Geol. Soc. Amer. Bull., 86, 39-46.

Isacks, B., Oliver, J., and Sykes, L. R., 1968. Seismology and the new global tectonics. J. Geophys. Res., 73, 5855-5899.

Ishihara, S., Igarashi, F., and Nishiwaki, C., 1975. A Re-examination of the regional distribution of the late Cenozoic ore deposits in the East Japan Arc. Geol. Soc. Amer. Bull., 86, 292-294.

Iturralde-Vinent, M. A., 1974. Circum-Caribbean tectonic and igneous activity and the evolution of the Caribbean plate: A discussion. Geol. Soc. Amer. Bull., 85, 1961-1962.

Jackson, H., 1973. Fluid transfer of heat in contact metamorphism. Geol. Soc. Amer. Abstr. Prog., 5, 325.

Jackson, E. D., and Shaw, H. R., 1975. Stress fields in central portions of the Pacific plate: Delineated in time by linear volcanic chains. J. Geophys. Res., 80, 1861-1874.

Jackson, E. D., Silver, E. A., and Dalrymple, G. B., 1972. Hawaiian Emperor chain and its relation to Cenozoic circumpacific tectonics. Geol. Soc. Amer. Bull., 83, 601-618.

Jacobs, J. A., 1963. The Earth's Core and Geomagnetism. Pergamon, Oxford, 137 pp.

Jacobs, J. A., Russell, R. D., and Wilson, J. T., 1959. Physics and Geology. McGraw-Hill, New York, 424 pp.

Jacoby, W. R., 1970. Instability in the upper mantle and global plate movements. J. Geophys. Res., 75, 5671-5680.

Jakes, P., 1973. Geochemistry and continental growth. In: Tarling and Runcorn (Editors) (q. v.), vol. 2, pp. 999-1009.

Jakes, P., and White, A. J. R., 1972. Major and trace element abundances in volcanic rocks of orogenic areas. Geol. Soc. Amer. Bull., 83, 29-40.

Jeffreys, H., 1962. The Earth: Origin, History, and Physical Constitution. Cambridge Univ. Press, Cambridge, 438 pp.

Jenks, W. F., 1975. Origins of some massive pyritic ore deposits of Western Europe. Econ. Geol., 70, 488-498.

Jerome, S. E., 1966. Some features pertinent in exploration of porphyry copper deposits. In: Titley and Hicks (Editors) (q. v.), pp. 75-86.

Johnson, L. R., 1967. Array measurements of velocities in the upper mantle. J. Geophys. Res., 72, 6309-6325.

Johnson, G. A. L., 1973. Closing of the Carboniferous sea in western Europe. In: Tarling and Runcorn (Editors) (q. v.), vol. 2, pp. 843-850.

Johnson, K. S., 1975. Permian copper shales of southwestern United States. In: Bartholome (Editor) (q. v.), pp. 383-393.

Johnson, H., and Smith, B. L. (Editors), 1970. The Megatectonics of Continents and Oceans. Rutgers, New Brunswick (N. J.), 284 pp.

Johnson, B. D., Lapoville, A., Larue, B. M., Henry, D. M., and Hawkins, L. V., 1974. Marine geophysical investigations in Melanesia. Trans. Amer. Geophys. Union, 55, 301.

Jolly, W. T., 1969. Behavior of Cu, Zn, and Ni during prehnite-pumpellyite rank metamorphism of the Keweewanan basalts, northern Michigan. Econ. Geol., 69, 1118-1125.

Jones, E. L., and Irwin, W. P., 1974. Southeastern Alaska--A displaced continental fragment? (Geol. Surv. Prof. Ppr. 800-B). U. S. Geological Survey, Washington, pp. B211-B217.

Jones, I. M., Hurst, V. J., and Walker, R. L., 1973. Strontium isotope composition and origin of the amphibolites of the Cartersville-Villa Rica District, Georgia. Geol. Soc. Amer. Abstr. Prog., 5, 407-408.

Jordan, T. H., and Frazer, L. N., 1975. Crustal and upper mantle structure from Sp phases. J. Geophys. Res., 80, 1504-1518.

Juan, V. C., 1956. Physiography and geology of Taiwan. Proc. Pac. Sci. Cong., Pac. Sci. Assoc., 8th Cong., 2, 281-312.

Juan, V. C., 1958. Continental rifting and igneous activities in the Neogene marginal geosynclines of Taiwan. Proc. Geol. Soc. China, 1, 27-35.

Juan, V. C., 1975. Tectonic evolution of Taiwan. Tectonophys., 26, 197-212.

Julian, B. R., 1970. Regional variations in upper mantle structure in North America. Trans. Amer. Geophys. Union, 51, 359.

Junta de Energia Nuclear, 1968. Structural and tectonic synthesis of the central uranium province of Portugal. In: Stemprok (Editor) (q. v.), pp. 71-84.

Justus, P. S., and Roper, P. J., 1973. Structural development of the Brevard Zone, Part 1: Theories of origin and tectonics. Geol. Soc. Amer. Abstr. Prog., 5, 408.

Kamen-Kaye, M., 1970. Geology and productivity of Persian Gulf synclinorium. Amer. Assoc. Petrol. Geol. Bull., 54, 2371-2394.

Kamen-Kaye, M., 1972. Persian Tethys and Indian Ocean. Amer. Assoc. Petrol. Geol. Bull., 56, 1984-1999.

Kanasewich, E. R., and Gutowski, P. R., 1975. Detailed seismic analysis of a lateral mantle inhomogeneity. Earth Planet. Sci. Ltr., 25, 379-384.

Kanasewich, E. R., Ellis, R. M., Chapman, C. H., and Gutowski, P. R., 1973. Seismic array evidence of a core boundary source for the Hawaiian linear volcanic chain. J. Geophys. Res., 78, 1361-1371.

Kanes, W. H., Saadi, M., Erlich, E., and Alem, A., 1973. Moroccan crustal response to continental drift. Science, 180, 950-952.

Karig, D. E., 1970. Ridges and basins of the Tonga-Kermadec island arc system. J. Geophys. Res., 75, 239-254.

Karig, D. E., 1971. Origin and development of marginal basins in the western Pacific. J. Geophys. Res., 76, 2542-2561.

Karig, D. E., 1972. Remnant arcs. Geol. Soc. Amer. Bull., 83, 1057-1068.

Karig, D. E., and Sharman, G. F., III, 1975. Subduction and accretion in trenches. Geol. Soc. Amer. Bull., 86, 377-389.

Katili, J. A., 1972. Structure and age of the Indonesian tin belt with special reference to Bangka. Tectonophys., 4, 403-418.

Katili, J. A., 1975. Volcanism and plate tectonics in the Indonesian island arcs. Tectonophys., 26, 165-188.

Katsumata, M., and Sykes, L. R., 1969. Seismicity and tectonics of the western Pacific: Izu-Mariana-Caroline and Ryukyu-Taiwan regions. J. Geophys. Res., 74, 5923-5948.

Katz, M. B., 1974. Paired metamorphic belts in Precambrian granulite rocks in Gondwanaland. Geology, 2, 237-241.

Kaula, W. M., 1972. Global gravity and tectonics. In: Robertson (Editor) (q. v.), pp. 386-405.

Kaula, W. M., 1975. Absolute plate motions by boundary velocity minimizations. J. Geophys. Res., 80, 244-248.

Kay, M., 1951. North American geosynclines. Geol. Soc. Amer. Mem. 48, 143 pp.

Keast, A., 1973. Contemporary biotas and the separation sequence of the southern continents. In: Tarling and Runcorn (Editors) (q. v.), vol. 1, pp. 309-343.

Keays, R. R., and Kirkland, M. C., 1972. Hydrothermal mobilization of gold from copper-nickel sulfides and ore genesis at the Thomson River Copper Mine, Victoria, Australia. Econ. Geol., 67, 1263-1275.

Keith, M. L., 1972. Ocean-floor convergence: a contrary view of global tectonics. J. Geol., 80, 249-276.

Kelleher, J., Sykes, L., and Oliver, J., 1973. Possible criteria for predicting earthquake locations and their application to major plate boundaries of the Pacific and Caribbean. J. Geophys. Res., 78, 2547-2585.

Kelly, W. C., and Clark, B. R., 1975. Sulfide deformation studies: III. Experimental deformation of chalcopyrite to 2,000 bars and 500°C. Econ. Geol., 70, 431-453.

Kemp, W. R., and Payne, A. L., 1975. Petrochemical associations within the Foothill Copper-Zinc Belt, Sierra Nevada, California. Geol. Soc. Amer. Abstr. Prog., 7, 332.

Kennedy, M. J., 1973. The relationship between pre-Ordovician polyphase deformation and ophiolite obduction in the Newfoundland Appalachian system. Geol. Soc. Amer. Abstr. Prog., 5, 183.

Kennison, J. E., 1966. The Mission copper deposit, Arizona. In: Titley and Hicks (Editors) (q. v.), pp. 281-287.

Kent, P. E., 1973. The continental margin of Tanzania. In: Tarling and Runcorn (Editors) (q. v.), vol. 2, pp. 949-952.

Kesler, S. E., 1973. Copper, molybdenum, and gold abundances in porphyry copper deposits. Econ. Geol., 68, 106-112.

Kesler, S. E., Jones, L. M., and Walker, R. L., 1975. Intrusive rocks associated with porphyry copper mineralization in island arc areas. Econ. Geol., 70, 515-526.

Khain, V. E., 1954. Geotektonicheskie osnovy poiskov nefti. Aznefteizdat, Baku (U. S. S. R.).

Khain, V. E., 1972. Main trends in the evolution of the Earth's crust (lithosphere). Internat. Geol. Cong. Abstr., 24, 84.

Khain, V. E., and Muratov, M. V., 1969. Crustal movements and tectonic structure of continents. In: Hart (Editor) (q. v.), pp. 523-538.

Kidd, W. S. F., Burke, K. C. A., and Wilson, J. T., 1974. Hot spots and the driving mechanism. Trans. Amer. Geophys. Union, 55, 439.

King, L. C., 1962. The Morphology of the Earth. Hafner, New York, 699 pp.

King, R. E. (Editor), 1972. Stratigraphic Oil and Gas Fields. Amer. Assoc. Petrol. Geol., Tulsa, 687 pp.

King, L. C., 1973. An improved reconstruction of Gondwana-land. In: Tarling and Runcorn (Editors) (q. v.), vol. 2, pp. 851-864.

King, J. R., Bruce, W. R., Slater, M. N., and Field, C. W., 1975. The geology and mineral deposits of the Cuddy Mountains, West Central Idaho. Geol. Soc. Amer. Abstr. Prog., 7, 617.

Kirkham, R. V., 1974. A synopsis of Canadian stratiform copper deposits in sedimentary sequences. In: Bartholome (Editor) (q. v.), pp. 367-382.

Kisvarsanyi, G., 1973. Geological sequence of igneous activity in the Precambrian St. Francois terrane, Missouri. Geol. Soc. Amer. Abstr. Prog., 5, 327.

Kisvarsanyi, E. B., 1974. Operation Basement: Buried Precambrian rocks of Missouri--their petrography and structure. Amer. Assoc. Petrol. Geol. Bull., 58, 674-684.

Klerkx, J., Deutsch, S., and De Paepe, 1974. Rubidium, strontium content and strontium isotope composition of strongly alkalic basaltic rocks from the Cape Verde islands. Contr. Mineral. Petrol., 45, 107-118.

Knopoff, L., 1964. The convection current hypothesis. Rev. Geophys., 2, 89-122.

Knopoff, L., 1967a. Thermal convection in the Earth's mantle. In: Gaskell (Editor) (q. v.), pp. 171-196.

Knopoff, L., 1967b. A convection in the upper mantle. Geophys. J., 14, 341-346.

Knopoff, L., 1969. Continental drift and convection. In: Hart (Editor) (q. v.), pp. 683-689.

Knopoff, L., and Fouda, A. A., 1975. Upper-mantle structure under the Arabian Peninsula. Tectonophys., 26, 121-134.

Knopoff, L., and Shapiro, J. N., 1969. Comments on the interrelationships between Gruneisen parameter and shock of isothermal equation of state. J. Geophys. Res., 74, 1439-1450.

Knopoff, L., Drake, D. L., and Hart, P. J. (Editors), 1968. The Crust and Upper Mantle of the Pacific Area. Amer. Geophys. Union, Washington, 522 pp.

Knowlton, F. H., 1919. Evolution of geologic climates. Geol. Soc. Amer. Bull., 30, 499-565.

Koo, J., and Mossman, D. J., 1975. Origin and metamorphism of the Flin Flon stratabound Cu-Zn sulfide deposit, Saskatchewan and Manitoba. Econ. Geol., 70, 48-62.

Koppen, W. P., and Wegener, A., 1924. Die Klimate der geologischen Vorzeit. Borntraeger, Berlin, 256 pp.

Krason, J., 1974. Central European versus south central U. S. A. geological settings of the Permian basins and associated copper mineralizations. Geol. Soc. Amer. Abstr. Prog., 6, 111.

Krauskopf, K. B., 1971. The source of ore metals. Geochim. Cosmochim. Acta, 35, 643-659.

Krogh, T. E., and Davis, G. L., 1972. The Grenville front

interpreted as an ancient plate boundary. Internat. Geol. Cong. Abstr., 24, 85.

Kroner, A., 1975. Proterozoic crustal evolution in parts of southern Africa and evidence for extensive sialic crust since the end of the Archean. Mtg. Roy. Soc. Lond., Mar. 13-14.

Kuenen, P. H., 1967. Geosynclinal sedimentation. Geologische Rundschau, 56, 1-19.

Kuno, H., 1959. Origin of Cenozoic petrographic provinces of Japan and surrounding areas. Bull. Volcanol., ser. 2, 20, 37-76.

Kurita, T., 1973. A procedure for elucidating fine structure of the crust and upper mantle for sedimentological data. Bull. Seis. Soc. Amer., 63, 189-209.

Kutina, J., 1975. Tectonic development and metallogeny of Madagascar with reference to the fracture pattern of the Indian Ocean. Geol. Soc. Amer. Bull., 86, 582-592.

Kutina, J., and Fabbri, A., 1972. Relationship of structural lineaments and mineral occurrences in the Abitibi area of the Canadian Shield (Paper 71-9). Geol. Surv. Canada, Ottawa, 36 pp.

Lachenbruch, A. H., 1973. A simple mechanical model for oceanic spreading centers. J. Geophys. Res., 78, 3395-3417.

Lachenbruch, A. H., and Thompson, G. A., 1972. Oceanic ridges and transform faults: their intersection angles and resistance to plate motion. Earth Planet. Sci. Ltr., 15, 116-122.

Lambert, I. B., and Sato, T., 1974. The Kuroko and associated ore deposits of Japan: A review of their features and metallogenesis. Econ. Geol., 69, 1215-1236.

Lambert, I. B., and Wyllie, P. J., 1970. Low-velocity zone of the Earth's mantle: incipient melting caused by water. Science, 196, 764-766.

Lamey, C. A., 1966. Metallic and Industrial Mineral Deposits. McGraw-Hill, New York, 567 pp.

Lange, I. M., and Wehrenberg, J. P., 1975. Precambrian belt

Revett and sandstone type red bed copper deposits of the southwestern United States. Geol. Soc. Amer. Abstr. Prog., 7, 339-340.

Lanphere, M. A., and Reed, B. L., 1973. Timing of Mesozoic and Cenozoic plutonic events in Circum-Pacific North America. Geol. Soc. Amer. Abstr. Prog., 5, 70.

Laraya, R. G., 1973. Interpretation of magnetic anomalies of plutonic intrusives in the search for porphyry copper deposits. Geol. Soc. Amer. Abstr. Prog., 5, 71

Larson, R. L., and Chase, C. G., 1970. Relative velocities of the Pacific, North America, and Cocos plates in the middle America region. Earth Planet. Sci. Ltr., 7, 425-428.

Laubscher, H., 1969. Mountain building. Tectonophys., 7, 551-563.

Laughton, A. S., Sclater, J. G., and McKenzie, D. P., 1973. The structure and evolution of the Indian Ocean. In: Tarling and Runcorn (Editors) (q. v.), vol. 1, pp. 203-212.

Laznicka, P., and Wilson, H. D. B., 1972. The significance of a copper-lead line in metallogeny. Internat. Geol. Cong. Abstr., 24, 136.

Lee, W. H. K. (Editor), 1965. Terrestrial Heat Flow. Amer. Geophys. Union, Washington, 276 pp.

Lee, W. H. K., and MacDonald, G. J. F., 1963. The global variation of terrestrial heat flow. J. Geophys. Res., 68, 6481-6492.

Leech, G. B., 1965. The Rocky Mountain Trench. Geol. Surv. Can. Ppr. 66-14, 307-329.

Lefebvre, J. J., 1974. Mineralisations cupro-cobaltiferes associees aux horizons pyroclastiques situes dans le faisceau superieur de la serie de Roan, a Sheturu, Shaba, Zaire. In: Bartholome (Editor) (q. v.), pp. 103-122.

Lehmann, E., 1972. On the source of the iron in the Lahn ore deposits. Min. Dep., 7, 247-270.

Leichti, P., 1960. The geology of Sarawak, Brunei, and the

western part of North Borneo. Geol. Surv. Dept. Brit. Terr. Borneo Bull., 3(1), 1-360.

Leitch, E. C., 1975. Plate tectonic interpretation of the Paleozoic history of the New England fold belt. Geol. Soc. Amer. Bull., 86, 141-144.

Le Pichon, X., 1968. Sea floor spreading and continental drift. J. Geophys. Res., 73, 3661-3697.

Le Pichon, X., 1969. Models and structure of oceanic crust. Tectonophys., 7, 385-401.

Le Pichon, X., Francheteau, J., and Bonnin, J., 1973. Plate Tectonics. Elsevier, Amsterdam, 300 pp.

Levinson, A. A., 1974. Introduction to Exploration Geochemistry. Applied Publishing, Calgary (Canada), 612 pp.

Levitte, D., Columba, J., and Mohr, P., 1974. Reconnaissance geology of the Amaro Horst, southern Ethiopia rift. Geol. Soc. Amer. Bull., 85, 417-422.

Lindgren, W., 1933. Mineral Deposits, 4th ed. McGraw-Hill, New York, 930 pp.

Liu, L., 1973. Chemistry of the Earth's lower mantle. J. Geophys. Res., 78, 3501-3504.

Lochman-Balk, C., 1974. Late Dresbachian (Late Cambrian) biostratigraphy of North America. Geol. Soc. Amer. Bull., 85, 135-140.

Lougheed, M. S., and Mancuso, J. J., 1973. Hematite framboids in the Negaunee iron formation, Michigan: Evidence for their biogenic origin. Econ. Geol., 68, 202-209.

Lowell, J. D., 1973. Porphyry copper deposits of the Southwest. Econ. Geol., 68, 139.

Lowell, J. D., 1974. Regional characteristics of porphyry copper deposits of the Southwest. Econ. Geol., 69, 601-617.

Lowell, J. D., and Genik, G. S., 1972. Sea-floor spreading and structural evolution of southern Red Sea. Amer. Assoc. Petrol. Geol. Bull., 56, 247-259.

Lowell, J. D., and Guilbert, J. M., 1970. Lateral and vertical alteration-mineralization zoning in porphyry ore deposits. Econ. Geol., 65, 373-408.

Lowell, G. R., and Sides, J. R., 1973. The occurrence and origin of rapakivi granite in the St. Francois Mountain batholith of southeast Missouri. Geol. Soc. Amer. Abstr. Prog., 5, 332-333.

Lowes, F. J., and Runcorn, S. K., 1951. The analysis of geomagnetic secular variation. Phil. Trans. Roy. Soc., Ser. A, 243, 525-535.

Lowry, W. D., 1974. North American geosynclines--Test of continental-drift theory. Amer. Assoc. Petrol. Geol. Bull., 58, 575-620.

Lubimova, E. A., 1969. Thermal history of the Earth. In: Hart (Editor) (q. v.), pp. 63-77.

Lubimova, E. A., and Nikitina, I. N., 1975. On heat flow singularities over mid-ocean ridges. J. Geophys. Res., 80, 232-243.

Lusk, J., 1972. Examination of volcanic-exhalative and biogenic origins for sulfur in the stratiform sulfide deposits of New Brunswick. Econ. Geol., 67, 169-183.

Lyustikh, E. N., 1969. Problem of convection in the Earth's mantle. In: Hart (Editor) (q. v.), pp. 689-692.

Macdonald, G. A., and Katsura, T., 1962. Relationship of petrographic suites in Hawaii. In: Macdonald and Kuno (Editors) (q. v.), pp. 187-195.

Macdonald, G. A., and Kuno, H. (Editors), 1962. The Crust of the Pacific Basin. Amer. Geophys. Union, Washington, 195 pp.

Macdonald, K., Luyendyk, B. P., Mudie, J. D., and Spiess, F. N., 1975. Near-bottom geophysical study of the Mid-Atlantic Ridge median valley near lat 37°N: Preliminary observations. Geology, 3, 211-215.

MacGillavry, H. J., 1970. Turbidite detritus and geosyncline history. Tectonophys., 9, 365-393.

MacIntyre, D. G., 1975. Volcanic-plutonic regimes and porphyry copper occurrences, Tahtsa Lake area, west central British Columbia. Geol. Soc. Amer. Abstr. Prog., 7, 811.

Madden, T. R., and Swift, C. M., 1969. Magnetotelluric studies of the electrical conductivity structure of the crust and upper mantle. In: Hart (Editor) (q. v.), pp. 469-479.

Malmquist, D., and Parasnis, D. S., 1972. Aitik: geophysical documentation of a third-generation copper deposit in north Sweden. Geoexplor., 10(3), 149-158.

Malpas, J., Stevens, R. K., and Strong, D. F., 1973. The origin and significance of ophiolite-amphibolite associations. Geol. Soc. Amer. Abstr. Prog., 5, 192.

Mammerickx, J., Anderson, R. N., Menard, H. W., and Smith, S. M., 1975. Morphology and tectonic evolution of the east-central Pacific. Geol. Soc. Amer. Bull., 86, 111-118.

Mannard, G. W., 1973. Syngenetic massive sulphide deposits. Econ. Geol., 68, 139.

Markowitz, W., and Guinot, B. (Editors), 1968. Continental Drift, Secular Motion of the Pole, and Rotation of the Earth. Reidel, Dordrecht, 107 pp.

Markowitz, W., Stoyko, N., Fedorov, E. P., 1964. Longitude and latitude. In: H. Odishaw (Editor), Research in Geophysics, vol. 2. MIT Press, Cambridge, pp. 149-162.

Marlow, M. S., Scholl, D. W., and Garrison, L. E., 1973. Comparative geologic histories of the Aleutian and Lesser Antilles island arcs. Geol. Soc. Amer. Abstr. Prog., 5, 77-78.

Martin, H., 1969. Problems of age relations and structure in some metamorphic belts of southern Africa. Geol. Assoc. Can. Spl. Ppr. 5, pp. 17-26.

Matsuda, T., and Uyeda, S., 1971. On the Pacific-type orogeny and its model--extension of the paired belt concept and possible origin of marginal seas. Tectonophys., 11, 5-27.

Matsumoto, T. (Editor), 1967. Age and Nature of the Circum-Pacific Orogenesis. Tectonophys., 4, 317-623.

Mattison, J. M., 1975. Early Paleozoic ophiolite complexes of Newfoundland: Isotopic ages of zircons. Geology, 3, 181-183.

Maucher, A., 1972. Time and stratabound ore deposits and the evolution of the Earth. Internat. Geol. Cong. Abstr., 24, 137.

Mavko, G., and Nur, A., 1975. Melt squirt in the asthenosphere. J. Geophys. Res., 80, 1444-1448.

Max, M. D., and Riddihough, R. P., 1975. Continuation of the Highland Boundary fault in Ireland. Geology, 3, 206-210.

Maxwell, J. C., 1968. Continental drift and a dynamic Earth. Amer. Sci., 56, 35-51.

Maxwell, A. E., 1969. Recent deep sea drilling results from the South Atlantic. Trans. Amer. Geophys. Union, 50, 113.

McBirney, A. R., 1969a. Compositional variation in Cenozoic calc-alkaline suites of Central America. Ore. Dept. Geol. Min. Res. Bull., 65, 185-189.

McBirney, A. R., 1969b. Andesitic and rhyolitic volcanism of orogenic belts. In: Hart (Editor) (q. v.), pp. 501-507.

McBirney, A. R., 1970. Cenozoic igneous events of the Circum-Pacific. Geol. Soc. Amer. Abstr. Prog., 2, 749-751.

McElhinny, M. W., 1972. Paleomagnetism and plate tectonics. Internat. Geol. Cong. Abstr., 24, 92.

McElhinny, M. W., 1973. Paleomagnetic results from Eurasia. In: Tarling and Runcorn (Editors) (q. v.), vol. 1, pp. 77-85.

McElhinny, M. W., Briden, J. C., Jones, D. L., and Brock, A., 1968. Geological and geophysical implications of paleomagnetic results from Africa. Rev. Geophys., 6, 201-238.

McGarr, A., 1969. Amplitude variations of Rayleigh waves: Propagation across a continental margin. Bull. Seis. Soc. Amer., 59, 1281-1305.

McGetchin, T. R., Nikhanj, Y. S., and Chodos, A. A., 1973.

Carbonate-kimberlite relations in the Cane Valley diatreme, San Juan County, Utah. J. Geophys. Res., 78, 1854-1869.

McGlynn, J. C., and Irving, E., 1974. Paleomagnetism of early Aphebian dibase dykes from the Slave Structural Province, Canada. Tectonophys., 26, 23-38.

McKenzie, D. P., 1969. Speculations on the causes and consequences of plate motions. Geophys. J., 18, 1-32.

McKenzie, D. P., 1972. Plate tectonics. In: Robertson (Editor) (q. v.), pp. 323-360.

McKenzie, D. P., and Morgan, W. J., 1969. The evolution of triple junctions. Nature, 224, 125-133.

McKenzie, D. P., and Parker, R. L., 1967. The north Pacific: an example of tectonics of a sphere. Nature, 216, 1276-1280.

McKerrow, W. S., and Ziegler, A. M., 1972. Paleozoic oceans. Nature Phys. Sci., 240, 92-94.

Melendres, M. M., and Comsti, F., 1951. Reconnaissance geology of southeastern Davao. Philippine Geol., 5, 38-46.

Melville, R., 1973. Continental drift and plant distribution. In: Tarling and Runcorn (Editors) (q. v.), vol. 1, pp. 439-446.

Menard, H. W., 1972. History of oceans basins. In: Robertson (Editor) (q. v.), pp. 440-460.

Menard, H. W., 1973. Epeirogeny and plate tectonics. Trans. Amer. Geophys. Union, 54, 1244-1255.

Merrill, R. B., and Wyllie, P. J., 1975. Kaersutite and kaersutite eclogite from Kakanui, New Zealand--Water-excess and water-deficient melting to 30 kilobars. Geol. Soc. Amer. Bull., 86, 555-570.

Meservey, R., 1969. Topological inconsistency of continental drift to the present-sized Earth. Science, 166, 609-611.

Metz, R. A., and Rose, A. W., 1966. Geology of the Ray copper deposit, Ray, Arizona. In: Titley and Hicks (Editors) (q. v.), pp. 177-188.

Meyerhoff, A. A., 1970. Continental drift: Implications of paleomagnetic studies, meteorology, physical oceanography, and climatology. J. Geol., 78, 1-51.

Meyerhoff, H. A., 1974. Sedimentary facies and plate tectonics of equatorial Pacific: Discussion. Amer. Assoc. Petrol. Geol. Bull., 58, 540-541.

Meyerhoff, A. A., and Teichert, C., 1971. Continental drift, III: Late Paleozoic glacial centers, and Devonian-Eocene coal distribution. J. Geol., 79, 285-321.

Milanovsky, E. E., 1972. Continental rift zones: their arrangement and development. Tectonophys., 15, 65-70.

Miller, R. J. M., 1973. The morphology of some massive sulfide deposits. Econ. Geol., 68, 139.

Milsom, J. S., 1970. Woodlark Basin, a minor center of sea-floor spreading in Melanesia. J. Geophys. Res., 75, 7335-7339.

Minear, J. W., and Toksoz, M. N., 1970. Thermal regime of a downgoing slab and new global tectonics. J. Geophys. Res., 75, 1397-1419.

Minoza, W. A., Ocampo, A. T., and Bellosillo, S., 1960. Significant Philippine earthquakes, 1949-1959. Philippine Weather Bur. Sci. Ppr. 101.

Misik, M. (Editor), 1968. Orogenic Belts. Academia, Prague, 327 pp.

Mitchell-Thome, R. C., 1972. Outline of the geology of the Cape Verde archipelago. Geol. Rund., 61, 1087-1108.

Mitchell, A. H. G., and Garson, M. S., 1972. Relationship of porphyry copper and circum-Pacific tin deposits to paleo-Benioff zones. Trans. Inst. Min. Metall., 81, 10-25.

Mitchell, A. H., and Reading, H. G., 1969. Continental margins, geosynclines, and ocean floor spreading. J. Geol., 77, 629-646.

Miyashiro, A., 1961. Evolution of metamorphic belts.

Tectonophys., 2, 277-311.

Miyashiro, A., 1972. Metamorphism and related magmatism in plate tectonics. Amer. J. Sci., 272, 656.

Miyashiro, A., 1973a. Metamorphism and Metamorphic Belts. Halsted, New York, 492 pp.

Miyashiro, A., 1973b. Paired and unpaired metamorphic belts. Tectonophys., 17, 241-254.

Miyashiro, A., 1973c. Paired and unpaired metamorphic belts. Trans. Amer. Geophys. Union, 54, 211.

Molnar, P., and Atwater, T., 1973. On the relative motion of "hotspots". Nature, 246, 288-290.

Molnar, P., and Oliver, J., 1969. Lateral variation of attenuation in the upper mantle and discontinuities in the lithosphere. J. Geophys. Res., 74, 2648-2682.

Monger, J. W. W., Souther, J. G., and Gabrielse, H., 1972. Evolution of the Canadian Cordillera: a plate-tectonic model. Amer. J. Sci., 272, 577-602.

Moore, G. D., 1973. Plate-edge deformation and crustal growth, Gulf of Mexico structural province. Geol. Soc. Amer. Bull., 84, 1883-1906.

Moore, W. J., and Nash, J. T., 1974. Alteration and fluid inclusion studies of the porphyry copper ore body at Bingham, Utah. Econ. Geol., 69, 631-645.

Moores, E. M., 1973. Plate tectonic significance of Alpine peridotite types. In: Tarling and Runcorn (Editors) (q. v.), vol. 2., pp. 963-975.

Morgan, W. J., 1968. Rises, trenches, great faults, and crustal blocks. J. Geophys. Res., 73, 1959-1982.

Morgan, W. J., 1970. Plate motions and deep mantle convection. Trans. Amer. Geophys. Union, 51, 822.

Morgan, W. J., 1971. Convection plumes in the lower mantle. Nature, 230, 42-43.

Morgan, W. J., 1972. Deep mantle convection plumes and plate motion. Amer. Assoc. Petrol. Geol. Bull., 56, 203-213.

Morioka, M., and Kigoshi, K., 1975. Lead isotopes and age of Hawaiian lherzolite nodules. Earth Planet. Sci. Ltr., 25, 116-120.

Morton, R. D., Goble, R. J., and Fritz, P., 1974. The mineralogy, sulfur-isotope composition, and origin of some copper deposits in the Belt Supergroup, southwest Alberta, Canada. Min. Dep. 9, 223-241.

Mouratov, M. V., 1972. Main structural elements of the crust on continents, their interrelations, and age. Internat. Geol. Cong. Abstr., 24, 93.

Muehlberger, W. R., 1971. Buried basement rocks of North America: their evidence for continental rifting and drifting. Geol. Soc. Amer. Abstr. Prog., 3, 653-654.

Muehlberger, W. R., and Ritchie, A. W., 1975. Caribbean-Americas plate boundary in Guatemala and southern Mexico as seen on Skylab IV orbital photography. Geology, 3, 232-235.

Muller, D. W., 1972. The geology of the Beltana willemite deposits. Econ. Geol., 67, 1146-1167.

Muller, S. C., 1975. Distribution and genesis of a copper mineralization near Pocatello, Idaho. Geol. Soc. Amer. Abstr. Prog., 7, 632-633.

Murray, C. G., 1970. Magma genesis and heat flow: Differences between mid-oceanic ridges and African rift valleys. Earth Planet. Sci. Ltr., 9, 34-38.

Myers, J. S., 1974. Cretaceous stratigraphy and structure, western Andes of Peru between latitudes 10° - $10^{o}30'$. Amer. Assoc. Petrol. Geol. Bull., 58, 474-487.

Naha, K., and Halyburton, R. V., 1974. Early Precambrian stratigraphy of central and southern Rajastan, India. Precam. Res., 1, 55-73.

Nairn, A. E. M., 1973. The argument for a southern Uratlantic.

In: Tarling and Runcorn (Editors) (q. v.), vol. 2, pp. 799-804.

Naldrett, A. J., 1970. Synvolcanic ultramafic bodies: a new class. Geol. Soc. Amer. Abstr. Prog., 2, 633-634.

Natland, J. H., 1974. Geology and tectonics of the Fagaloa shield volcano, island of Upolo, Western Somoa: Plate bending, faulting, and possible shear melting north of the Tonga trench. Geol. Soc. Amer. Abstr. Prog., 6, 226.

Nelson, T. H., and Temple, P. G., 1972. Mainstream mantle convection: A geologic analysis of plate motion. Amer. Assoc. Petrol. Geol. Bull., 56, 226-246.

Newhouse, W. H. (Editor), 1942. Ore Deposits as Related to Structural Features. Princeton Univ. Press, Princeton (N. J.), 280 pp.

Nikolayev, A. V., and Tregub, F. S., 1970. A statistical model of the Earth's crust: Method and results. Tectonophys., 10, 573-578.

Nitsan, U., 1973. Viscous heat production in a slab. J. Geophys. Res., 78, 1395-1397.

Noble, J. A., 1974. Metal provinces and metal finding in the western United States. Min. Dep., 9, 1-25.

Normand, D. E., 1973. Precambrian geology of the northern Sangre de Cristo Range, Chaffee, Fremont, and Saguache Counties, Colorado. Geol. Soc. Amer. Abstr. Prog., 5, 499.

North, F. K., 1971. Alpine serpentinites, ocean ridges, and continental drift. Geol. Mag., 108, 81-102.

Norton, S. A., 1973. Laterite and bauxite formation. Econ. Geol., 68, 353-361.

Ocola, L. C., and Meyer, R. P., 1970. Regional upper crustal structure of mid-continent of U. S. A. Geol. Soc. Amer. Abstr. Prog., 2, 638.

Ohmoto, H., 1972. Systematics of sulfur and carbon isotopes in hydrothermal ore deposits. Econ. Geol., 67, 551-578.

Ohmoto, H., and Rye, R. O., 1974. Hydrogen and oxygen isotopic compositons of fluid inclusions in the Kuroko deposits, Japan. Econ. Geol., 69, 947-953.

Olade, M. A., and Fletcher, W. K., 1975. Primary dispersion of rubidium and strontium around porphyry copper deposits, Highland Valley, British Columbia. Econ. Geol., 70, 15-21.

Oliver, J., and Isacks, B., 1967. Deep earthquake zones, anomalous structures in the upper mantle, and the lithosphere. J. Geophys. Res., 72, 4259-4275.

Oliver, J., Sykes, L., and Isacks, B., 1969. Seismology and the new global tectonics. Tectonophys., 7, 527-547.

Ollier, C., 1970. Volcanoes. MIT Press, Cambridge, 177 pp.

Orajaka, S. O., 1973. Possible metallogenic provinces in Nigeria. Econ. Geol., 68, 278-280.

Orlin, H. (Editor), 1966. Gravity Anomalies: Unsurveyed Areas. Amer. Geophys. Union, Washington, 142 pp.

Osmaston, M. F., 1973. Limited lithosphere separation as a main cause of continental basins, continental growth, and epeirogeny. In: Tarling and Runcorn (Editors) (q. v.), vol. 2, pp. 649-674.

Ostenso, N. D., 1973. Sea-floor spreading and the origin of the Arctic basin. In: Tarling and Runcorn (Editors) (q. v.), vol. 1, pp. 165-173.

Oxburgh, E. R., 1972. Plate tectonics and continental collision. Nature, 239, 202-204.

Oyarzun, M. J., 1975. Porphyry copper and tin-bearing porphyries: A discussion of genetic models. Phys. Earth Planet. Int., 9, 259-263.

Page, R. W., and McDougall, I., 1972a. Geochronology of the Panguna porphyry copper deposit, Bougainville Island, New Guinea. Econ. Geol., 67, 1065-1074.

Page, R. W., and McDougall, I., 1972b. Ages of mineralization of gold and porphyry copper deposits in the New Guinea highlands. Econ. Geol., 67, 1034-1048.

Palmer, A. R., 1974. Search for the Cambrian world. Amer. Sci., 62, 216-224.

Papa, A., 1972. Conceptions nouvelles sur la structure des Albanides. Bull. Soc. Geo. France, ser. 7, 12, 1096-1109.

Parak, T., 1973. Rare earths in the apatite iron ore of Lappland together with some data about the Sr, Th, and U content of these ores. Econ. Geol., 68, 210-221.

Park, C. F., 1972. The iron ore deposits of the Pacific basin. Econ. Geol., 67, 339-349.

Parkhurst, R. J., 1974. Rb-Sr whole-rock chronology of Caledonian events in northeast Scotland. Geol. Soc. Amer. Bull., 85, 345-370.

Pautot, G., Renard, V., Daniel, J., and Dupont, J., 1973. Morphology, limits, origin, and age of salt layer along South Atlantic African margin. Amer. Assoc. Petrol. Geol. Bull., 57, 1658-1671.

Peive, A. V., Perfiliev, A. S., and Ruzhentsev, S. V., 1972. Problems in intracontinental geosynclines. Internat. Geol. Cong. Abstr., 24, 95.

Pelissonnier, H., 1968. Sur la distribution du cuivre economiquement exploitable sur la Terre. In: Ahrens (Editor) (q. v.), pp. 1139-1149.

Pelissonnier, H., 1972. Les Dimensions des Gisements de Cuivre du Monde. Bu. Recherches Geol. et Min., Paris, 405 pp.

Pereira, J., and Dixon, C. J., 1971. Mineralization and plate tectonics. Min. Dep., 6, 404-405.

Peryt, T. M., 1975. Genesis of evaporite-associated stratiform metalliferous deposits: A sabkha process. Econ. Geol., 70, 407-409.

Peters, W. C., 1973. Characteristics of the Alpine-

Mediterranean copper deposits. Econ. Geol., 68, 140.

Petrascheck, W. E., 1973. Some aspects of the relations between continental drift and metallogenic provinces. In: Tarling and Runcorn (Editors) (q. v.), vol. 1, pp. 567-572.

Petrushevsky, B. A., 1969. Earthquakes and tectonics. In: Hart (Editor) (q. v.), pp. 279-282.

Pettijohn, F. J., 1970. The Canadian Shield--A status report, 1970. In: Baer (Editor) (q. v.), pp. 239-265.

Phillips, C. H., Gambell, N. A., and Fountain, D. S., 1974. Hydrothermal alteration, mineralization, and zoning in the Ray deposit. Econ. Geol., 69, 1237-1250.

Piper, J. D. A., 1973a. Geological interpretation of paleomagnetic results from the African Precambrian. In: Tarling and Runcorn (Editors) (q. v.), vol.1, pp. 19-31.

Piper, J. D. A., 1973b. The geology of constructive plate margins. In: Tarling and Runcorn (Editors) (q. v.), vol. 2, pp. 625-634.

Piper, J. D. A., 1975a. Paleomagnetic correlations of Precambrian formations of east-central Africa and their tectonic implications. Tectonophys., 26, 135-161.

Piper, J. D. A., 1975b. Paleomagnetic evidence for a Proterozoic super-continent. Mtg. Roy. Soc. Lond., Mar. 13-14.

Piper, J. D. A., 1975c. Paleomagnetism of Silurian lavas of Somerset and Gloucestershire, England. Earth Planet. Sci. Ltr., 25, 355-360.

Pittman, W. C., III, and Talwani, M., 1972. Sea floor spreading in the North Atlantic. Geol. Soc. Amer. Bull., 83, 619-646.

Poldervaart, A. (Editor), 1955. Crust of the Earth. Geol. Soc. Amer., Boulder, 761 pp.

Polyak, B. G., and Smirnov, Ya. B., 1966. Heat flow on continents. Dokl. Acad. Nauk SSSR, 168, 170-172.

Pouba, Z., and Stemprok, M. (Editors), 1970. Problems of

Hydrothermal Ore Deposition. Schweitzerbatsche Verlag, Stuttgart, 396 pp.

Press, F., 1972. The Earth's interior as inferred from a family of models. In: Robertson (Editor) (q. v.), pp. 147-171.

Preto, V. A., Lefebure, D. V., and Atkinson, S. J., 1975. The Nicola Group: Mesozoic volcanism related to rifting in southern British Columbia. Geol. Soc. Amer. Abstr. Prog., 7, 841-842.

Price, R. H., and Mountjoy, E. W., 1970. Geologic structure of the Canadian Rocky Mountains between Bow and Athabasca rivers: A progress report. Geol. Assoc. Can. Spl. Ppr., 6, 7-25.

Purser, B. H. (Editor), 1973. The Persian Gulf. Springer-Verlag, New York, 471 pp.

Purvis, A. C., Nesbit, R. W., and Halberg, J. A., 1972. The geology of part of Carr Boyd Rocks Complex and its associated nickel mineralization, Western Australia. Econ. Geol., 67, 1093-1113.

Pushcharovsky, Y. M., and Arkhipov, I. V., 1972. Tectonic aspects of the Chilean Andes. Geotectonics, 1, 44-52.

Putzer, H., 1972. Kupferschiefer in Zambia und Mitteleuropa, ein palaogeographischer Vergleich. Geol. Jahrb., 90, 415-430.

Qamar, A., 1974. Seismicity of the Baffin Bay region. Bull. Seis. Soc. Amer., 64, 87-98.

Qureshy, M. N., Venkatachalam, S., and Subrahmanyan, C., 1974. Vertical tectonics in the Middle Himalayas: An appraisal from recent gravity data. Geol. Soc. Amer. Bull., 85, 921-926.

Radkevich, E. A., 1972. Metallogenic zonality of the Pacific ore belt. Internat. Geol. Cong. Abstr., 24, 144.

Radulescu, D. P., and Sandulescu, M., 1973. The plate-tectonics concept and the geological structure of the Carpathians. Tectonophys., 16, 155-173.

Raitt, R. W., Shor, G. C., Francis, T. J. G., and Morris, G. B., 1969. Anisotropy of the Pacific upper mantle. J. Geophys. Res., 74, 3095-3109.

Ramberg, I. B., and Smithson, S. B., 1975. Gridded fault patterns in a later Cenozoic and Paleozoic continental rift. Geology, 3, 201-205.

Rastogi, B. K., 1974. Earthquake mechanisms and plate tectonics in the Himalayan region. Tectonophys., 21, 47-56.

Rehrig, W. H., and Heidrick, T. L., 1972. Regional fracturing in Laramide stocks of Arizona and its relationship to porphyry copper mineralization. Econ. Geol., 67, 198-213.

Reinhard, M., and Wenk, E., 1951. Geology of the Colony of North Borneo. Geol. Sur. Dept. Brit. Terr., Borneo Bull., 1, 1-160.

Renfro, A. R., 1974. Genesis of evaporite-associated stratiform metalliferous deposits: A sabkha process. Econ. Geol., 69, 33-45.

Rentzsch, J., 1975. The Kupferschiefer in comparison with the deposits of the Zambian copperbelt. In: Bartholome (Editor) (q. v.), pp. 395-418.

Reyment, R. A., 1973. Cretaceous history of the South Atlantic Ocean. In: Tarling and Runcorn (Editors) (q. v.), vol. 2, pp. 805-814.

Richards, J. R., 1971. Major lead orebodies--mantle origin? Econ. Geol., 66, 425-434.

Rickard, D. T., 1974. Low temperature copper geochemistry: Gitological aspects. In: Bartholome (Editor) (q. v.), pp. 1-34.

Rickard, D. T., and Zweifel, H., 1975. Genesis of Precambrian sulfide ores, Skellefte district, Sweden. Econ. Geol., 70, 255-274.

Ridge, J. D., 1973. Volcanic exhalation and ore deposition in the vicinity of the sea floor. Econ. Geol., 68, 140.

Riding, R., 1974. Model of the Hercynian foldbelt. Earth Planet. Sci. Ltr., 24, 125-135.

Rikitake, T., 1966. Electromagnetism of the Earth's Interior. Elsevier, Amsterdam, 308 pp.

Ringwood, A. E., 1974. The petrological evolution of island

arc systems. J. Geol. Soc. Lond., 130, 183-204.

Ritsema, A. R., and Veldkamp, J., 1960. Fault plane mechanisms of southeast Asian earthquakes. Koninkl. Ned. Meteorol. Inst. Mededel. en Verh., 76.

Rittmann, A., 1962. Volcanoes and Their Activity. Wiley, New York, 305 pp.

Roaldset, E., 1975. Rare earth element distributions in some Precambrian rocks and their phyllosilicates, Numedal, Norway. Geochim. Cosmochim. Acta, 39, 455-469.

Roberts, R. J., 1972. Evolution of the Cordilleran fold belt. Geol. Soc. Amer. Bull., 83, 1989-2004.

Roberts, R. G., 1975. The geological setting of the Mattagami Lake mine, Quebec: A volcanogenic massive sulfide deposit. Econ. Geol., 70, 115-129.

Robertson, E. C. (Editor), 1972. The Nature of the Solid Earth. McGraw-Hill, New York, 677 pp.

Robertson, D., 1973. Plate tectonism and the search for petroleum and base metals. Austral. Min., 65(8), 72-74.

Robinson, P. L., 1973. Paleoclimatology and continental drift. In: Tarling and Runcorn (Editors) (q. v.), vol. 1, pp. 451-476.

Robinson, R. F., and Cook, A., 1966. The Safford copper deposit, Lone Star mining district, Graham County, Arizona. In: Titley and Hicks (Editors) (q. v.), pp. 251-266.

Rockwell, D. W., and Roberts, R., 1973. Operation for application-oriented exploration data-base systems shown. Oil Gas J., 71(8), 62-78.

Roeder, D. H., 1973. Subduction and orogeny. J. Geophys. Res., 78, 5005-5024.

Rogers, J. J. W., and 7 others, 1974. Paleozoic and lower Mesozoic volcanism and continental growth in the western United States. Geol. Soc. Amer. Bull., 85, 1913-1924.

Romer, A. S., 1973. Vertebrates and continental connections: An introduction. In: Tarling and Runcorn (Editors) (q. v.), vol. 1, pp. 345-349.

Rona, P. A., and Dietz, R. S., 1974. Sediment accumulation, sea-floor spreading, and eustacy: Reply. Geol. Soc. Amer. Bull., 85, 831-832.

Ronov, A. B., Mikhaylovskaya, M. S., Solodkova, I. I., 1965. The evolution of the chemical and mineral composition of sandy rocks. Geochem. Internat., 2, 318-371.

Roper, P. J., 1974. Plate tectonics: A plastic as opposed to a rigid body model. Geology, 2, 247-250.

Rose, A. W., and Baltosser, W. W., 1966. The porphyry copper deposit at Santa Rita, New Mexico. In: Titley and Hicks (Editors) (q. v.), pp. 205-220.

Ross, D. A., and Schlee, J., 1973. Shallow structure and geologic development of the southern Red Sea. Geol. Soc. Amer. Bull., 84, 3827-3848.

Rowlands, N. J., 1974. The gitology of some Adelaidean stratiform copper occurrences. In: Bartholome (Editor) (q. v.), pp. 419-427.

Roy, R. F., Blackwell, D. D., and Decker, E. R., 1972. Continental heat flow. In: Robertson (Editor) (q. v.), pp. 506-543.

Ruitenberg, A. A., Chandra, J., and Ruitenberg, G., 1972. Metallization in the Caledonian belt, southern New Brunswick. Can. Min. Metallurg. Bull., 65(725), 81-90.

Runcorn, S. K. (Editor), 1962. Continental Drift. Academic, New York, 338 pp.

Runcorn, S. K., 1969. Convection in the mantle. In: Hart (Editor) (q. v.), pp. 692-698.

Runcorn, S. K., 1973. Polar wandering and continental drift. In: Tarling and Runcorn (Editors) (q. v.), vol. 2, pp. 995-997.

Runcorn, S. K., 1974. On the forces not moving lithospheric

plates. Tectonophys., 21, 197-202.

Rundkvist, D. V., 1968. Accumulation of metals and the evolution of the genetic types of deposits in the history of the Earth. In: Stemprok (Editor) (q. v.), pp. 85-97.

Russell, J. R., 1972. Baseline studies on the effects of low magnetic fields (less than 50 gammas) on development processes on Fundulus heteroclitus. Alabama Acad. Sci. J., April.

Russell, M. J., 1973. Base-metal mineralization in Ireland and Scotland and the formation of the Rockall Trough. In: Tarling and Runcorn (Editors) (q. v.), vol. 1, pp. 581-597.

Rutland, R. W. R., 1973. Tectonic evolution of the continental crust of Australia. In: Tarling and Runcorn (Editors) (q. v.), vol. 2, pp. 1011-1033.

Rye, R. O., and Ohmoto, H., 1974. Sulfur and carbon isotopes and ore genesis: A review. Econ. Geol., 69, 826-842.

Sacks, I. S., and Okada, H., 1975. A comparison of the anelasticity structure beneath western South America and Japan. Phys. Earth Planet. Int., 9, 211-219.

Saegart, W. E., Sell, J. D., and Kilpatrick, B. E., 1974. Geology and mineralization of La Caridad porphyry copper deposit, Sonora, Mexico. Econ. Geol., 69, 1060-1077.

Sakai, H., and Matsubaya, O., 1974. Isotopic geochemistry of Japan and its bearing on the Kuroko ore solutions. Econ. Geol., 69, 974-991.

Salop, L. I., and Scheinmann, Yu. M., 1969. Tectonic history and structure of platforms and shields. Tectonophys., 7, 565-597.

Sammis, C., Jordan, T., and Anderson, D. L., 1970. Inhomogeneity in the upper mantle. Trans. Amer. Geophys. Union, 51, 828.

Sangster, D. F., 1972. Precambrian volcanogenic massive sulphide deposits in Canada: A review. Geol. Surv. Can. Spl. Ppr. 72-2, 44 pp.

Santos, V., 1969. Preliminary report on the geology and

mineral resources of central Palawan. Philippine Geol., 13(4), 104-141.

Sarkar, S. C., and Deb, M., 1969. Metamorphism of sulfides of the Singhbhum copper belt, India: The evidence from the ore belt. Econ. Geol., 69, 1282-1293.

Sato, T., 1972. Metallogenesis in Neogene base metal and Au-Ag deposits in Japan; with special reference to kuroko deposits. Min. Geol., 4, 185-204.

Sato, K., and Sasaki, A., 1973. Lead isotopes of the black ore ("kuroko") deposits from Japan. Econ. Geol., 68, 547-552.

Sawkins, F. J., 1972. Sulfide ore deposits in relation to plate tectonics. J. Geol., 80, 377-397.

Saxena, M. W., 1972. Validity of continental drift vis-a-vis the Indian shield and the Himalayas. Internat. Geol. Cong. Abstr., 24, 98.

Schau, M., 1975. Meta-komatiites in a stable Archean environment. Geol. Soc. Amer. Abstr. Prog., 7, 849.

Scheidegger, A. E., 1963. Principles of Geodynamics. Academic, New York, 362 pp.

Scheidegger, A. E., 1972. The rheology of the tectonosphere. Internat. Geol. Cong. Abstr., 24, 99.

Scheidegger, K. F., 1973. Temperatures and compositions of magmas ascending along mid-ocean ridges. J. Geophys. Res., 78, 3340-3355.

Schenk, P. E., 1971. Southeastern Atlantic Canada, northwestern Africa, and continental drift. Can. J. Earth Sci., 8, 1218-1251.

Schenk, P. E., 1973. The early Paleozoic of Nova Scotia with analogues and plate tectonics. Geol. Soc. Amer. Abstr. Prog., 5, 216.

Schilling, J. G., 1975a. Azores mantle blob: Rare-earth evidence. Earth Planet. Sci. Ltr., 25, 103-115.

Schilling, J. G., 1975b. Rare-earth variations across "normal segments" of the Reykjanes Ridge, 60°-53^oN, Mid-Atlantic Ridge, 29^oS, and East Pacific Rise, 2°-19^oS, and evidence on the composition of the underlying low-velocity layer. J. Geophys. Res., 80, 1459-1473.

Schilling, J. G., and Bonatti, E., 1975. East Pacific Ridge (2^oS-19^oS) versus Nazca intraplate volcanism: Rare-earth evidence. Earth Planet. Sci. Ltr., 25, 93-102.

Schlanger, S. O., and 10 others, 1974. Testing a hot-spot theory. Geotimes, 19(3), 16-19.

Schmitt, H. A., 1966. The porphyry copper deposits in their regional setting. In: Titley and Hicks (Editors) (q. v.), pp. 17-34.

Schneider, E. D., 1972. Sedimentary evolution of rifted continental margins. Geol. Soc. Amer. Mem. 132, 109-118.

Scholl, D. W., and Marlow, M. S., 1974. Sedimentary sequences in modern Pacific trenches and the deformed circum-Pacific eugeosyncline. In: Dott and Shaver (Editors) (q. v.), pp. 193-211.

Schopf, J. M., 1973. Coal, climate, and global tectonics. In: Tarling and Runcorn (Editors) (q. v.), vol. 1, pp. 609-622.

Schuchert, C., 1923. Sites and natures of North American geosynclines. Geol. Soc. Amer. Bull., 34, 151-260.

Schwarts, G. M., 1966. The nature of primary and secondary mineralization in porphyry copper deposits. In: Titley and Hicks (Editors) (q. v.), pp. 41-50.

Schwarzer, R. R., and Rogers, J. W., 1973. Alkali olivine basalt series of the Davis Mountains, South Central Texas: A comparison with worldwide occurrences. Geol. Soc. Amer. Abstr. Prog., 5, 102.

Schwarzer, R. R., and Rogers, J. J. W., 1974. A worldwide comparison of alkali olivine basalts and their differentiation trends. Earth Planet. Sci. Ltr., 23, 286-296.

Schwarzschild, M., 1958. Structure and Evolution of the Stars. Princeton Univ. Press, Princeton, 296 pp.

Sclater, J. G., and Fisher, R. E., 1974. Evolution of the East Central Indian Ocean, with emphasis on the tectonic setting of the Ninetyeast Ridge. Geol. Soc. Amer. Bull., 85, 683-702.

Sclater, J. G., and Francheteau, J., 1970. The implications of terrestrial heat flow observations on current tectonic and geochemical models of the crust and upper mantle of the Earth. Geophys. J., 20, 493-509.

Sclater, J. G., and McKenzie, D. P., 1973. Paleobathymetry of the south Atlantic. Geol. Soc. Amer. Bull., 84, 3203-3216.

Sclater, J. G., Anderson, R. N., and Bell, M. L., 1971a. Elevation of ridges and evolution of the central Pacific. J. Geophys. Res., 76, 7888-7915.

Sclater, J. G., Hawkins, J. W., and Chase, C. G., 1971b. Crustal extension between the Tonga and Lau ridges: Petrologic and geophysical evidence. Trans. Amer. Geophys. Union, 52, 194.

Sclater, J. G., Lawver, L. A., and Parsons, B., 1975. Comparison of long-wavelength residual elevation and free air gravity anomalies in the North Atlantic and possible implications for the thickness of the lithosphere plate. J. Geophys. Res., 80, 1031-1052.

Scott, M. R., Scott, R. B., Rona, P. A., Butler, L. W., and Nalwalk, A. J., 1974. Rapidly accumulating manganese deposit from the median valley of the Mid-Atlantic Ridge. Geophys. Res. Ltr., 1, 355-358.

Scott, H. S., 1965. Marine seismic exploration in the Philippines. Philippine Geol., 19(2), 51-56.

Scrutton, R. A., 1973. The age relationship of igneous activity and continental break-up. Geol. Mag., 110, 227-234.

Seeger, C. R., 1975. Absolute motion of the Eurasian plate: A problem in vector geometry: Comment. Geology, 3, 159.

Seward, T. M., 1973. Thio complexes of gold and the transport of gold in hydrothermal ore solutions. Geochim. Cosmochim. Acta, 37, 379-399.

Shackleton, R. M., 1973. Correlation of structures across Precambrian orogenic belts in Africa. In: Tarling and Runcorn (Editors), (q. v.), vol. 2, pp. 1091-1095.

Shackleton, R. M., 1975. Pan African structures. Mtg. Roy. Soc. Lond., Mar. 13-14.

Shank, R. J., 1972. Copper. Can. Min. J., 93(2), 79-89.

Shaw, H. R., 1973. Mantle convection and volcanic periodicity in the Pacific: Evidence from Hawaii. Geol. Soc. Amer. Bull., 84, 1505-1526.

Shaw, H. R., and Jackson, E. D., 1973. Linear island chains in the Pacific: Result of thermal plumes or gravitational anchors? J. Geophys. Res., 78, 8634-8652.

Shchlegov, A. D., 1968. Endogenous deposits of autonomous activization. In: Stemprok (Editor) (q. v.), pp. 43-55.

Shemyakin, V. M., 1972. An early Precambrian formation of intrusive charnokites found in the Baltian shield. Proc. USSR Acad. Sci., Geol., 9, 40-45.

Shimer, J. A., 1968. The Changing Earth. Harper and Row, New York, 233 pp.

Shipulin, F. K., 1972. On the formation of metal-bearing solutions in deep magmatic bodies. Internat. Geol. Cong. Abstr., 24, 148.

Shtreys, N. A., and Tseysler, V. M., 1972. Formation of geosynclinal regions. Geotectonics, 1, 1-2.

Shults, S. S., 1972. Planetary fractures and tectonic deformations. Geotectonics, 4, 203-207.

Silbering, N. J., 1973. Geologic events during the Permo-Triassic time along the Pacific margin of the United States. The Permian and Triassic Systems and the Mutual Boundary. Alta. Soc.

Petrol. Geol., Calgary, pp. 345-362.

Sillitoe, R. H., 1972a. A plate tectonic model for the origin of porphyry copper deposits. Econ. Geol., 67, 184-197.

Sillitoe, R. H., 1972b. Relation of metal provinces in western America to subduction of oceanic lithosphere. Geol. Soc. Amer. Bull., 83, 813-818.

Sillitoe, R. H., 1972c. Formation of certain massive sulphide deposits at sites of sea-floor spreading. Trans. Inst. Min. Metall., 81, B141-148.

Sillitoe, R. H., 1973. Geology of the Los Pelambres porphyry copper deposit, Chile. Econ. Geol., 68, 1-10.

Silver, E. A., von Huene, R., and Crouch, J. K., 1974. Tectonic significance of the Kodiak-Bowie seamount chain, north-eastern Pacific. Geology, 2, 147-150.

Simmons, W. W., and Fowells, J. E., 1966. Geology of the Copper Cities mine. In: Titley and Hicks (Editors) (q. v.), pp. 151-156.

Simonen, A., 1969. Batholiths and their orogenic setting. In: Hart (Editor) (q. v.), pp. 488-489.

Sipkin, S. A., and Jordan, T. H., 1975. Lateral heterogeneity of the upper mantle determined from the travel times of ScS. J. Geophys. Res., 80, 1474-1484.

Smirnov, V. I., 1972. The relations between syngenetic and epigenetic processes during formation of stratiform ore deposits in USSR. Internat. Geol. Cong. Abstr., 24, 149-150.

Smith, W. D., 1952. Geology and mineral resources of the Philippine Islands. Bu. Sci. Publ. 19, Manila.

Smith, A. G., 1972a. Estimation of tectonic rotation poles of inactive structures. Geol. Soc. Amer. Mem. 132, 23-33.

Smith, A. G., 1972b. Pre-Mesozoic plate tectonics. Internat. Geol. Cong. Abstr., 24, 100.

Smith, A. G., 1973. The so-called Tethyan ophiolites. In:

Tarling and Runcorn (Editors) (q. v.), vol. 2, pp. 977-986.

Smith, J. W., Schopf, J. W., and Caplan, I. R., 1970. Extractable organic matter in Precambrian cherts. Geochim. Cosmochim. Acta, 34, 659-675.

Sollogub, V. B., 1969. Seismic crustal studies in southeastern Europe. In: Hart (Editor) (q. v.), pp. 189-195.

Sollogub, V. B., 1970. On certain regularities of crustal structure associated with the major geologic features of southeastern Europe. Tectonophys., 10, 549-559.

Solomon, M., Graves, D. I., and Klominsky, J., 1972. Metallogenic provinces and districts in the Tasman orogenic zone of eastern Australia. Proc. Austral. Inst. Min. Metall., 242, 9-24.

Solomon, M., and Griffiths, J. R., 1972. Tectonic evolution of the Tasman orogenic zone, eastern Australia. Nature Phys. Sci., 237, 3-6.

Solomon, S. C., Sleep, N. H., and Richardson, R. M., 1974. Driving forces, absolute plate motions, and intraplate stress. Trans. Amer. Geophys. Union, 55, 438.

Sopko, P. F., 1972. Tectonomagmatic and metallogenic features of pyrite-bearing ore districts. Internat. Geol. Cong. Abstr., 24, 151-152.

Spall, H., 1972. Paleomagnetism and Precambrian continental drift. Internat. Geol. Cong. Abstr., 24, 101.

Spence, C. D., 1975. Volcanogenic features of the Vauze sulfide deposit, Noranda, Quebec. Econ. Geol., 70, 102-114.

Spooner, J., Williams, L., Kennedy, A., and Burcher, M., 1974. Mining Annual Review. Mining Journal, London, 506 pp.

Spurr, J. E., 1923. The Ore Magmas, vol. 2. McGraw-Hill, New York.

Stacey, F. D., 1969. Physics of the Earth. Wiley, New York, 324 pp.

St. Amand, P., 1959. Circum-Pacific orogeny. Publ. Domin.

Obs. Ottawa, 20, 403-411.

St. Amand, P., 1961. Observaciones e interpretacion de los terremotos chilenos de 1960. Univ. Chile Communic. Esc. Geol., 1(2).

St. Amand, P., and Allen, C. R., 1960. Strike-slip faulting in northern Chile. Geol. Soc. Amer. Bull., 71, 1965.

Stanton, R. L., 1972. Ore Petrology. McGraw-Hill, New York, 713 pp.

Steinhart, J. S., and Smith, T. J. (Editors), 1966. The Earth beneath the Continents. Amer. Geophys. Union, Washington, 663 pp.

Steinhorn, T. L., 1973. Depth-of-origin of magmas of the high cascades as inferred from geochemical relations. Geol. Soc. Amer. Abstr. Prog., 5, 110.

Steinmann, G., 1905. Geologische Beobachtungen in den Alpen. Ber. Natur. Ges., Freiburg, 1(16), 44-65.

Stemprok, M. (Editor), 1968. Endogenous Ore Deposits. Academia, Prague, 425 pp.

Stepanov, P. I., and Mironov, S. I., 1937. Geologiya mestorozhdenii kaustobiolitov. Otdel Nauchno-Tekhnicheskoi Informatsii, Moscow.

Stephansson, O., 1975. Polydiapirism of granitic rocks in the Svecofennian of central Sweden. Precamb. Res., 2, 189-214.

Stevens, R. K., Strong, D. F., and Kean, B. F., 1974. Do some eastern Appalachian ultramafic rocks represent mantle diapirs produced above a subduction zone? Geology, 2, 175-178.

Stevenson, J. S. (Editor), 1962. The Tectonics of the Canadian Shield. Univ. Toronto Press, 182 pp.

Stewart, J. H., 1972. Initial deposits in the Cordilleran geosyncline: evidence of a late Precambrian (850 m. y.) continental separation. Geol. Soc. Amer. Bull., 83, 1345-1360.

Stewart, J. H., and Poole, F. G., 1975. Extension of the Cordilleran miogeosynclinal belt to the San Andreas fault, Southern

California. Geol. Soc. Amer. Bull., 86, 205-212.

Stewart, J. H., Walker, G. W., and Kleinkampl, F. J., 1975. Oregon-Nevada lineament. Geology, 3, 265-268.

Stille, H., 1940. Einfuehrung in den Bau Nordamerikas. Bontraeger Verlagsbuchhandlung, Berlin, 717 pp.

Stringham, B., 1966. Igneous rock types and host rocks associated with porphyry copper deposits. In: Titley and Hicks (Editors) (q. v.), pp. 35-40.

Strong, D. F., 1973. Volcanic, intrusive, and tectonic activity in an Ordovician island arc of central Newfoundland. Geol. Soc. Amer. Abstr. Prog., 5, 224.

Subbotin, S. I., Sollogub, V.B., Khain, V. Y., Slavin, V. I., and Chekunov, A. V., 1972. Some remarks concerning the structure and evolution of the Earth's crust. Hung. Geophys. Inst. Geophys. Trans., 1972, 151-165.

Sugimura, A., 1960. Zonal arrangement of some geophysical and petrological features in Japan and its environs. Univ. Tok. Fac. Sci. J., ser. 2, 12, 133-153.

Sugimura, A., 1968. Spatial relation of basaltic magmas in island arcs. In: H. H. Hess and A. Poldervaart (Editors), Basalts: The Poldervaart Treatise on Rocks of Basaltic Composition, vol. 2. Interscience, New York, pp. 537-571.

Sugimura, A., and Uyeda, S., 1973. Island Arcs: Japan and Its Environs. Elsevier, Amsterdam, 247 pp.

Sutherland-Brown, A., Cathro, R. J., Panteleyev, A., and Ney, C. S., 1971. Metallogeny of the Canadian Cordillera. Can. Inst. Min. Metall. Trans., 74, 121-145.

Sutton, J., 1963. Long-term cycles in the evolution of the continents. Nature, 198, 731-735.

Sutton, J., 1970. Migration of high temperature zones in the crust. In: S. K. Runcorn (Editor), Paleogeophysics. Academic, New York, pp. 365-375.

Sutton, J., 1973. Some changes in continental structure since early Precambrian time. In: Tarling and Runcorn (Editors) (q. v.), vol. 2, pp. 1071-1081.

Sutton, J., Shackleton, R. M., and Briden, J. C., 1975. Global tectonics in Proterozoic times. Mtg. Roy. Soc. Lond., Mar. 13-14.

Sutulov, A., 1973. *Mineral Resources and the Economy of the USSR*. McGraw-Hill, New York, 192 pp.

Sutulov, A., 1974. *Copper Porphyries*. Univ. Utah. Salt Lake City, 200 pp.

Sychev. P. M., 1973. Upper-mantle structure and nature of deep processes in island arcs and trench systems. *Tectonophys.*, *19*, 349-359.

Sylvester-Bradley, P. C., 1968. Tethys, the lost ocean. *Sci. J.*, *4*, 47-53.

Takeuchi, H., Uyeda, S., and Kanamori, H., 1967. *Debate about the Earth*. Freeman Cooper, San Francisco, 253 pp.

Tappan, H., and Loeblich, A. R., 1970. Geobiologic implications of fossil phytoplankton evolution and time-space distribution. *Geol. Soc. Amer. Spl. Ppr. 127*, 247-340.

Tarling, D. H., 1973. Metallic ore deposits and continental drift. *Nature*, *243*, 193-196.

Tarling, D. H., and Runcorn, S. K. (Editors), 1973. *Implications of Continental Drift to the Earth Sciences*, 2 vols. Academic, New York, 1184 pp.

Tatsch, J. H., 1959. Initial results of preliminary investigations (1938-1958) regarding the feasibility of using polyistic planetary configurations for making certain solar-system evolutionary studies. U. S. Embassy, Caracas, Venezuela, in manus.

Tatsch, J. H., 1960. The evolution of the solar system in accordance with a dual primeval planet model. U. S. Embassy, Caracas, Venezuela, private publication, given limited distribution to approximately 300 selected Earth and planetary scientists in various parts of the world.

Tatsch, J. H., 1962. The evolution of the retrograde motion of certain planetary satellites in accordance with a dual primeval planet model, outline of doctoral dissertation, submitted to graduate advisor, February 4, 1962, about 75 pp.

Tatsch, J. H., 1963a. Certain selenophysical implications of applying a dual primeval planet model to the Earth. _Trans_. _Amer_. _Geophys_. _Union_, _44_, 877.

Tatsch, J. H., 1963b. Certain seismological implications of applying a dual primeval planet model to the Earth. _Trans_. _Amer_. _Geophys_. _Union_, _44_, 887.

Tatsch, J. H., 1963c. Certian volcanological implications of applying a dual primeval planet model to the Earth. _Trans_. _Amer_. _Geophys_. _Union_, _44_, 892.

Tatsch, J. H., 1964a. Distribution of active volcanoes: Summary of preliminary results of three-dimentional least-squares analysis. _Geol_. _Soc_. _Amer_. _Bull_., _75_, 751-752.

Tatsch, J. H., 1964b. Certain geomagnetic implications of applying a dual primeval planet model to the Earth. _Trans_. _Amer_. _Geophys_. _Union_, _45_, 39.

Tatsch, J. H., 1964c. Certain oceanographical implications of applying a dual primeval planet model to the Earth. _Trans_. _Amer_. _Geophys_. _Union_, _45_, 74.

Tatsch, J. H., 1964d. Further volcanological implications of applying a dual primeval planet model to the Earth. _Trans_. _Amer_. _Geophys_. _Union_, _45_, 124.

Tatsch, J. H., 1965a. The global distribution patterns of fossil volcanoes and their relationship to the global distribution oftterns of active volcanoes. National Science Foundation proposal number P51073R.

Tatsch, J. H., 1966a. Certain correlations between geomagnetic observations in the Indian Ocean and deductions arrived at from applying a dual primeval planet model to that region of the Earth. Trans. Amer. Geophys. Union, 47, 56-57.

Tatsch, J. H., 1966b. Certain correlations between oceanographic observations along the South Atlantic Ridge and deductions arrived at from applying a dual priveval planet model to that region of the Earth. Trans. Amer. Geophys. Union, 47, 123.

Tatsch, J. H., 1966c. Certain correlations between geomagnetic observations in the South Atlantic Ocean and deductions arrived at from applying a dual primeval planet model to that region of the Earth. Trans. Amer. Geophys. Union, 47, 464.

Tatsch, J. H., 1966d. Certain correlations between oceanographic observations along the North Atlantic Ridge and deductions made from applying a dual primeval planet model to that region of the Earth. Trans. Amer. Geophys. Union, 47, 477.

Tatsch, J. H., 1966e. Certain correlations between selenophysical observations and deductions arrived at from applying a dual primeval planet model to the Earth-Moon system. Trans. Amer. Geophys. Union, 47, 486.

Tatsch, J. H., 1966f. Certain correlations between seismological observations in the African continent and deductions made from applying a dual primeval planet model to that region of the Earth. Trans. Amer. Geophys. Union, 47, 490.

Tatsch, J. H., 1967. Global geomagnetic evidence supporting the existence of a geophysical equator predicted as a consequence of applying a dual primeval planet model to the Earth. Trans. Amer. Geophys. Union, 48, 59.

Tatsch, J. H., 1969. Sea-floor spreading, continental drift, and plate tectonics unified into a single global concept by the application of a dual primeval planet hypothesis to the Earth. Trans. Amer. Geophys. Union, 50, 672.

Tatsch, J. H., 1970. Global seismicity patterns as interpreted in accordance with a dual primeval planet hypothesis. Geol. Soc. Amer. Abstr. Prog., 2, 153.

Tatsch, J. H., 1972a. The Earth's Tectonosphere: Its Past Development and Present Behavior. Tatsch Associates, Sudbury, 889 pp.

Tatsch, J. H., 1972b. Global symmetry of mineralization belts. Geol. Soc. Amer. Abstr. Prog., 4, 729-730.

Tatsch, J. H., 1973a. Mineral Deposits: Origin, Evolution, and Present Characteristics. Tatsch Associates, Sudbury, 264 pp.

Tatsch, J. H., 1973b. The origin, evolution, and present characteristics of the Tertiary petroleum deposits of the world. Tatsch Associates, Sudbury, 58 pp.

Tatsch, J. H., 1973c. The origin, evolution, and present characteristics of the Mesozoic petroleum deposits of the world. Tatsch Associates, Sudbury, 62 pp.

Tatsch, J. H., 1973d. The origin, evolution, and present characteristics of the Paleozoic petroleum deposits of the world. Tatsch Associates, Sudbury, 63 pp.

Tatsch, J. H., 1973e. The origin, evolution, and present characteristics of the giant oil fields of North America. Tatsch Associates, Sudbury, 67 pp.

Tatsch, J. H., 1973f. The origin, evolution, and present characteristics of the petroleum deposits in the Arctic Ocean and North Sea areas. Tatsch Associates, Sudbury, 62 pp.

Tatsch, J. H., 1973g. The origin, evolution, and present characteristics of the petroleum deposits on the Middle East areas. Tatsch Associates, Sudbury, 62 pp.

Tatsch, J. H., 1973h. The origin, evolution, and present characteristics of the petroleum deposits within the Indonesian islands and within other areas of the arc extending roughly from the Bay of Bengal to the Kermadec-Tonga region. Tatsch Associates, Sudbury, 66 pp.

Tatsch, J. H., 1973i. The origin, evolution, and present characteristics of the Tertiary basins that appear to be largely nonpetroliferous. Tatsch Associates, Sudbury, 62 pp.

Tatsch, J. H., 1973j. The probability for the existence of Tertiary petroleum deposits between Cape Cod and Cape Fear. Tatsch Associates, Sudbury, 65 pp.

Tatsch, J. H., 1973k. Petroleum in the Philippines: An analysis of related Mesozoic and Cenozoic geophenomena. Tatsch Associates, Sudbury, 43 pp.

Tatsch, J. H., 1974a. The Moon: Its Past Development and Present Behavior. Tatsch Associates, Sudbury, 338 pp.

Tatsch, J. H., 1974b. Petroleum Deposits: Origin, Evolution, and Present Characteristics. Tatsch Associates, Sudbury, 378 pp.

Tatsch, J. H., 1975a. Gold Deposits: Origin, Evolution, and Present Characteristics. Tatsch Associates, Sudbury, about 360 pp., in press.

Tatsch, J. H., 1975b. The Physics of the Solar System: An Analysis Based on the Dual Primeval Planet Hypothesis. Tatsch Associates, Sudbury, about 450 pp., in prep.

Tatsch, J. H., 1975c. The Principle of Universal Heterogeneity: The Basis for the Dual Primeval Planet Hypothesis. Tatsch Associates, Sudbury, about 475 pp., in prep.

Tatsch, J. H., 1975d. The Earth's Tectonosphere: Its Past Development and Present Behavior, 2nd ed. Tatsch Assoicates, Sudbury, about 550 pp., in press.

Tatsch, J. H., and Talbot, J. L., 1971. Global symmetry of tectonic belts. Geol. Soc. Amer. Abstr. Prog., 3, 728-729.

Tatsumoto, M., 1969. Lead isotopes in volcanic rocks and possible ocean-floor thrusting beneath island arcs. Earth Planet. Sci. Ltrs., 6, 369-376.

Tatsumoto, M., and Knight, J. J., 1969. Isotopic compositions of lead in volcanic rocks from central Honshu - with regard to basalt genesis. Geochem. J., 3, 53-86.

Taylor, R. G., 1972. The relationship between structure and ore-body type in the Ardlethan tinfield, N. S. W. Econ. Geol., 67, 116-118.

Taylor, R. G., and Steveson, B. G., 1972. An analysis of metal distribution and zoning in the Herberton tinfield, north Queensland, Australia. Econ. Geol., 67, 1234-1240.

Terman, M. J., 1974. Pre-Mesozoic plate tectonics of the Far East. Geol. Soc. Amer. Abstr. Prog., 6, 982.

Termier, H., and Termier, G., 1973. The early climate history of the Earth and the implications of "plate tectonics" in the context of organic evolution. In: Tarling and Runcorn (Editors) (q. v.), vol. 1, pp. 477-485.

Theodore, T. G., and Nash, J. T., 1973. Geochemical and fluid zonation of Copper Canyon, Lander County, Arizona. Econ. Geol., 68, 565-570.

Thompson, G., and Melson, W. G., 1972. The petrology of oceanic crust across fracture zones in the Atlantic Ocean: Evidence of a new kind of sea-floor spreading. J. Geol., 80, 526-532.

Thurlow, J. G., Swanson, E. A., and Strong, D. F., 1975. Geology and lithogeochemistry of the Buchons polymetallic sulfide deposits, Newfoundland. Econ. Geol., 70, 130-144.

Tikhonov, A. N., Lubimova, Ye. A., and Vsalov, V. K., 1969. Heat flow from the Earth's interior depending on inner parameters variations. Bull. Volcanol., 33(1), 261-280.

Tilling, R. I., Gottfried, D., and Rowe, J. J., 1973. Gold abundance in igneous rocks: Bearing on gold mineralization. Econ. Geol. 68, 168-186.

Titley, S. R., 1972. Intrusion and wall rock porphyry copper deposits. Econ. Geol., 67, 122-123.

Titley, S. R., 1973. Geological environment and characteristics of some porphyry copper deposits of the southwestern Pacific. Econ. Geol., 68, 141.

Titley, S. R., 1975. Geological characteristics and environment of some porphyry copper occurrences in the southwestern Pacific. Econ. Geol., 70, 499-514

Titley, S. R., and Hicks, C. L. (Editors), 1966. Geology of the Porphyry Copper Deposits. Univ. Ariz., Tucson, 287 pp.

Tozer, D. C., 1965. Heat transfer and convection currents. Phil. Trans. Roy. Soc. Lond., Ser. A, 258, 252-271

Trexler, D. W., 1974. Fold structures in northwestern Colorado from ERTS-1 imagery. Geol. Soc. Amer. Abstr. Prog., 6, 480.

Trofimov, V. S., 1972. On the origin of diamondiferous ecologite. Internat. Geol. Cong. Abstr., 24, 153.

Truchan, M., and Larson, R. L., 1973. Tectonic lineaments on the Cocos plate. Earth Planet. Sci. Ltr., 17, 426-433

Tschanz, C. M., Marvin, R. F., Cruz-B., J., Mehnert, H. H., and Cebula, G. T., 1974. Geologic evolution of the Sierra Nevada de Santa Marta, northwestern Colombia. Geol. Soc. Amer. Bull., 85, 273-284.

Tseysler, V. M., 1972. Relationship between Mesozoic and Paleogene formations in the platform and geosynclinal troughs of the Mediterranean belt in the southern USSR. Geotectonics, 1, 26-31.

Turcotte, D. L., and Oxburgh, E. R., 1972. Mantle convection and the new global tectonics. Ann. Rev. Fluid Mech., 4, 33-68.

Turcotte, D. L., and Oxburgh, E. R., 1974. The driving mechanism for plate tectonics and the state of stress in the lithosphere. Trans. Amer. Geophys. Union, 55, 438.

Turekian, K. K., 1968. The composition of the crust. In: Ahrens (Editor) (q. v.), pp. 551-557.

Umbgrove, J. H. F., 1956. Origin of continental shelves. Amer. Assoc. Petrol. Geol. Bull., 30, 249-253.

Upadhyay, H. D., and Strong, D. F., 1973. Geological setting of the Betts Cove copper deposits, Newfoundland: An example of ophiolite sulfide mineralization. Econ. Geol., 68, 161-167.

Uyeda, S., and Ben-Avraham, Z., 1972. Origin and development of the Philippine Sea. Nature, Phys. Sci., 240, 176-178.

Valentine, J. W., and Moores, E. M., 1972. Global tectonics and the fossil record. J. Geol, 80, 167-184.

Van Bemmelan, R. W., 1949. The Geology of Indonesia. Government Printing Office, The Hague.

Van Bemmelen, R. W., 1974. Drving forces of orogeny, with emphasis on blue-schist facies of metamorphism (Test case III: The Japan Arc). Tectonophys., 22, 83-125.

Van Breemen, O., Aftalion, M., Allaart, J. H., 1974. Isotopic and geochronomogical studies on granties from the Ketilidian mobile belt of south Greenland. Geol. Soc. Amer. Bull., 85, 403-412.

Van Krevelen, D. W., 1961. Coal. Elsevier, Amsterdam.

Van Schmus, W. R., 1975. Middle Precambrian rocks of the Great Lakes area, North America. Mtg. Roy. Soc. Lond., Mar. 13-14.

Veizer, J., 1975. 87-Sr/86-Sr composition of seawater during geologic history and its significance as an index of crustal evolution. Geol. Soc. Amer. Abstr. Prog., 7, 874.

Velasco, J. R., 1966. Geology of the Cananea district. In: Titley and Hicks (Editors) (q. v.), pp. 245-250.

Vening Meinesz, O. A., 1950. Les graben africaine resultat de compression ou de tension dans la croute terrestre? Bull. Inst. Roy. Col. Belg., 21, 539-552.

Vening Meinesz, F. A., 1954. Indonesian archipelago: A geophysical study. Bull. Geol. Soc. Amer., 65, 143-144.

Vening Meinesz, F. A., 1964. The Earth's Crust and Upper Mantle. Elsevier, Amsterdam, 124 pp.

Verhoogen, J., 1961.. Heat balance of the Earth's core. Geophys. J., Roy. Astr. Soc., 4, 276.

Verhoogen, J., 1965. Phase change in convection in the Earth's mantle. Phil. Trans. Roy. Soc. Lond., Ser. A., 258, 276-283.

Verma, R. K., and Mital, G. S., 1974. Paleomagnetic study of a vertical sequence of traps from Mount Pavagarh, Gujrat, India. Phys. Earth Planet. Int., 8, 75-82.

Vine, F. J., 1969a. Spreading of the ocean floor: New evidence. Science, 154, 1405-1415.

Vine, F. J., 1969b. Sea-floor spreading and continental drift. Geol. Soc. Amer. Abstr. Prog., 7, 231-232.

Vine, F. J., and Matthews, D. H., 1963. Magnetic anomalies over ocean ridges. Nature, 199, 947-949.

Vogel, D. E., 1973. Evidence for polymetamorphism at the Superior-Grenville boundary, south-central Quebec. Geol. Soc. Amer. Abstr. Prog., 5, 232-233.

Vogt, P. R., 1974. Volcano spacing, fractures, and thickness of the lithosphere. Earth Planet. Sci. Ltr., 21, 235-252.

Vogt, P. R., 1975. Changes in geomagnetic reversal frequency at times of tectonic change: Evidence for coupling between core and upper mantle processes. Earth Planet. Sci. Ltr., 25, 313-321.

Vogt, P. R., and Johnson, G. L., 1975. Transform faults and longitudi flow below the Midoceanic ridge. J. Geophys. Res., 80, 1399-1428.

Volborth, A., 1968. On the distribution and role of oxygen in the geochemistry of the Earth's crust. In: Ahrens (Editor) (q. v.), pp. 825-854.

Volkes, F. M., 1973. Metallogeny possibly related to continental break-up in southwest Scandinavia. In: Tarling and Runcorn (Editors) (q. v.), vol. 1, pp. 573-579.

Von Drascke, R., 1887. Fragmente zu einer Geologie der Insel Luzon (Philippinen). Gerolds Sohn, Vienna.

von Gehlen, K., 1968. Potential criteria for distinguishing between ores from crustal and sub-crustal sources. In: Stemprok

(Editor) (q. v.), pp. 179-184.

Vostokova, E. A., Marinov, N. A., and Clerova, L. I., 1972. Hydrological zonation of the African continent. Proc. USSR Acad. Sci., 2, 102-134.

Walcott, R. I., 1972. Gravity, flexure, and the growth of basins at the continental edge. Geol. Soc. Amer. Bull., 83, 1845-1848.

Walper, J. L., and Rowett, C. L., 1972. Plate tectonics and the origin of the Caribbean Sea and the Gulf of Mexico. Gulf Coast Assoc. Geol. Socs. Trans., 22, 105-116.

Wardlaw, N. C., 1972. Unusual marine evaporites with salts of calcium and magnesium chloride in Cretaceous basins of Sergipe, Brazil. Econ. Geol, 67, 156-168.

Warren, K., 1973. Mineral Resources. Halsted, New York, 272 pp.

Waterman, G. C., and Hamilton, R. L., 1975. The Sar Cheshmeh porphyry copper deposit. Econ. Geol., 70, 568-576.

Watson, B. N., 1973a. Post-laramide emplacement of porphyry copper deposits. Econ. Geol., 68, 283-284.

Watson, J., 1973b. Influence of crustal evolution on ore deposition. Inst. Min. Metall. Trans., B-82, 107-113.

Watson, J. V., 1975. Vertical movements in Proterozoic structural provinces. Mtg. Roy. Soc. Lond., Mar. 13-14.

Watterson, J., 1975. Proterozoic deformation patterns in Greenland. Mtg. Roy. Soc. Lond., Mar. 20-21.

Watts, A. B., 1974. Gravity anomalies landward of island arcs and the evolution of inter-arc basins. Trans. Amer. Geophys. Union, 55, 301.

Watts, A. B., and Cochran, J. R., 1974. Gravity anomalies in the Galapagos Islands area. Science, 184, 808-809.

Watts, A. B., and Talwani, M., 1975. Gravity effect of down-going lithospheric slabs beneath island arcs. Geol. Soc. Amer.

Bull., 86, 1-4.

Weaver, P., 1950. Variations in history of continental shelves. Amer. Assoc. Petrol. Geol. Bull., 34, 351-360.

Weber, W., and Stephenson, J. F., 1973. The content of mercury and gold in some Archean rocks of the Rice Lake (Manitoba) area. Econ. Geol., 68,401-407.

Wedepohl, K. H., 1971. Kupferschiefer as a prototype of syngenetic sedimentary ore deposits. Japan Soc. Min. Geol., SP-3, 268-273.

Wells, A. J., 1969. The crush zone of the Iranian Zagros Mountains and its implications. Geol. Mag., 106, 385-414.

Wegener, A., 1915. Die Entstehung der Kontinente und Ozeane. Sammlung Vieweg, Braunschweig, 94 pp.

Wensink, H., 1973. The India-Pakistan subcontinent and the Gondwana reconstructions based on paleomagnetic results. In: Tarling and Runcorn (Editors) (q. v.), vol. 1, pp. 103-116.

Westerman, D. S., 1973. Geological history and age relations of some igneous and metamorphic rocks in southeastern Maine. Geol. Soc. Amer. Abstr. Prog., 5, 236-237.

White, D. E., 1968a. Environments of generation of some base-metal ore deposits. Econ. Geol., 63, 301-335.

White, W. S., 1968b. The native-copper deposits of northern Michigan. Amer. Inst. Min. Metall. Petr. Engrs., 1, 303-325.

White, R. D., 1972. The Cumberland Basin: A possible rift. Can. Min. Metall. Bull., 65, 83-88.

White, D. E., 1974. Diverse origins of hydrothermal ore fluids. Econ. Geol., 69, 954-973.

Whitehead, E. V., 1970. Molecular evidence for the biogenesis of petroleum and natural gas. Internat. Symp. Hydrochem. Biochem., Tokyo.

Whitehead, J. A., Jr., and Luther, D. A., 1975. Dynamics of laboratory diapir and plume models. J. Geophys. Res., 80, 705-717.

Whiteman, A., Naylor, D., Pegrum, R., and Rees, G., 1975. North Sea troughs and plate tectonics. Tectonophys., 26, 39-54.

Whitney, J. A., 1975a. Vapor generation in a quartz monzonite magma: A synthetic model with application to porphyry copper deposits. Econ. Geol., 70, 346-358.

Whitney, J. W., 1975b. A resource analysis based on porphyry copper deposits and the cumulative copper metal curve using Monte Carlo simulation. Econ. Geol., 70, 527-537.

Wielchowsky, C. C., and Gilbert, O. E., 1973. The pre-Mesozoic rocks of the southeastern Alabama - South Georgia - Florida subsurface and their bearing on the origin of the southernmost Appalachians. Geol. Soc. Amer. Abstr. Prog., 5, 452.

Wiese, R. G., 1960. Petrology and geochemistry of a copper-bearing Precambrian shale, White Pine, Michigan, PhD dissertation, Harvard Univ.

Wiese, R. G., 1973. Mineralogy and geochemistry of the Parting Shale, White Pine, Michigan. Econ. Geol., 68, 317-331.

Wiggins, R. A., and Helmberger, D. V., 1973. Upper mantle structure of the western United States. J. Geophys. Res., 78, 1870-1880.

Wildeman, T. R., and Haskin, L. A., 1973. Rare earths in Precambrian sediments. Geochim. Cosmochim. Acta, 37, 419-438.

Willis, B., 1937. Geologic observations in the Philippine archipelago. Natl. Res. Cncl. Philippines Bull., 13.

Willis, B., 1944. Philippine earthquakes and structure. Bull. Seismol. Soc. Amer., 34, 69-81.

Wilson, J. T., 1954. The development and structure of the crust. In: G. P. Kuiper (Editor), The Earth as a Planet. Univ. Chicago Press, Chicago.

Wilson, R. A. M., 1961. The geology and mineral resources of the Banggi Island and Sugut River area, North Borneo. Geol. Sur. Dept. Brit. Terr. Borneo Mem. 15, 1-143.

Wilson, J. T., 1963. A possible origin of the Hawaiian Islands. Can. J. Phys., 41, 863-870.

Wilson, J. T., 1965. Convection currents and continental drift: evidence from ocean islands suggesting movement in the Earth. Phil. Trans. Roy. Soc., Ser. A., 258, 145-155.

Wilson, J. T., 1967. Theories of building of continents. In: Gaskell (Editor) (q. v.), pp. 445-473.

Wilson, J. T., 1969. Aspects of the different mechanics of ocean floors and continents. Tectonophys., 8, 281-284.

Wilson, C. J. L., 1972a. The stratigraphic and metamorphic sequence eest of Mt. Isa and associated igneous intrusions. Proc. Austral. Inst. Min. Metall., 243, 27-42.

Wilson, J. T., 1972b. New insights into old shields. Tectonophysics, 13, 73-94.

Wilson, M. R., 1973. The geological setting of the Sulitjelma ore bodies, central Norwegian Caledonides. Econ. Geol., 68, 307-316.

Wilson, R. C. L., 1975. Atlantic opening and Mesozoic continental margin basins of Iberia. Earth Planet. Sci. Ltr., 25, 33-43.

Wilson, H. D. B., and Laznicka, P., 1972. Copper belts, lead belts, and copper-lead lines of the world. Internat. Geol. Cong. Abstr., 24, 156.

Wiltse, M. A., 1973. Fe-Cu massive sulfide deposits in ancient outer-arc ridge-trench slope environments. Geol. Soc. Amer. Abstr. Prog., 5, 122-123.

Winterer, E. L., 1973. Sedimentary facies and plate tectonics of equatorial Pacific. Amer. Assoc. Petrol. Geol. Bull., 57, 265-282.

Winterer, E. L., 1974. Sedimentary facies and plate tectonics of equatorial Pacific: Reply. Amer. Assoc. Petrol. Geol. Bull., 58, 541-544.

Wissink, A., 1972. Les gisements de manganese du monde: Conditions du depot, typologie, et metal contenu. Geologie Appliquee, 11(1), 34-43.

Wolfe, J. A., 1973. Porphyry copper deposits in the Philippines. Econ. Geol., 68, 142.

Woodward, L. A., 1972. Upper Precambrian strata of the eastern Great Basin as host rocks for mineralization. Econ. Geol., 67, 677-681.

Woollard, G. P., 1969a. Tectonic activity in North America as indicated by earthquakes. In: Hart (Editor) (q. v.), pp. 125-133.

Woollard, G. P., 1969b. Standardization of gravity measurements. Ibid., pp. 283-293.

Woollard, G. P., 1969c. Regional variations in gravity. Ibid., pp. 320-341.

Woollard, G. P., 1972. Regional variation in gravity. In: Robertson (Editor) (q. v.), pp. 463-505.

Woolnough, W. G., 1946. Distribution of oceans and continents: A suggestion. Amer. Assoc. Petrol. Geol. Bull., 30, 1981-2012.

Wright, C., 1975. Comments on "Seismic array evidence of a core boundary source for the Hawaiian linear volcanic chain" by E. R. Kanesewich et al. J. Geophys. Res., 80, 1915-1919.

Wright, J. B., and McCurry, P., 1973. Sea-floor spreading and continental ore deposits. In: Tarling and Runcorn (Editors) (q. v.), vol. 1, pp. 563-565.

Wright, N. A., and Williams, P. L., 1974. Mineral Resources of Antarctica. U. S. Geol. Survey, Washingtoo, 29 pp.

Wyllie, P. J. (Editor), 1967. Ultramafic and Related Rocks. Wiley, New York, 464 pp.

Wyllie, P. J. (Editor), 1969a. The origin of ultramafic and ultrabasic rocks. Tectonophys., 7, 437-455.

Wyllie, P. J., 1969b. The ultramafic belts. In: Hart (Editor) (q. v.), pp. 480-488.

Wyllie, P. J., 1971. The Dynamic Earth: Textbook in Geosciences. Wiley, New York, 416 pp.

Wynne-Edwards, H. R. (Editor), 1969. Age Relations in High-Grade Metamorphic Terrains. Geol. Soc. Can., Toronto, 228 pp.

Wynne-Edwards, H. R., and Hasan, Z., 1970. Intersecting orogenic belts across the North Atlantic. Amer. J. Sci., 268, 289-308.

Yakovlev, G. F., and Starostin, V. I., 1972. Tectonophysical analysis of the sulphide ore-bearing fields. Internat. Geol. Cong. Abstr., 24, 157.

Yasui, M., Kishii, T., Watanabe, T., and Uyeda, S., 1968. Heat flow in the Sea of Japan. Geophys. Monogr., 12, 3-16.

Yoder, H. S., and Tilley, C. E., 1957. Basalt magmas. Carnegie Inst. Washington Yearbook, 56, 156-161.

York, J. E., and Helmberger, D. V., 1973. Low-velocity zone variations in the southwestern United States. J. Geophys. Res., 78, 1883-1886.

Yoshii, T., 1975. Regionality of group velocities of Rayleigh waves in the Pacific and thickening of the plate. Earth Planet. Sci. Ltr., 25, 305-312.

Young, G. M., 1974. Stratigraph8, paleocurrents, and stromatolites of Hadrynian (Upper Precambrian) rocks of Victoria Island, Arctic Archipelago, Canada. Precam. Res., 1, 13-41.

Ypma, P. J. M., 1972. The multistage emplacement of the Ni-Co-Bi-Ag deposits of the Chalanches, France: remobilization, rejuvenation, or repetition of ore-genetic patterns. Internat. Geol. Cong. Abstr., 24, 157-158

Zapletal, K., 1968. Tektogene der Erde. In: M. Mistik (Editor), Orogenic Belts. Academia, Prague, pp. 213-224.

Zen, E., 1972. The Taconide Zone and the Taconic Orogeny in

the Western Part of the Northern Appalachian Orogen (GSA Spl Ppr 135). Geol. Soc. Amer., Boulder, 72 pp.

Zijderweld, J. D. A., and Van der Voo, R., 1973. Paleomagnetism in the Mediterranean area. In: Tarling and Runcorn (Editors) (q. v.), vol. 1, pp. 133-161.

Zlobin, B. I., Belyaev, Y. I., 1970. On the probability of primary or secondary nature of copper source in hydrothermal polymetallic ores of the central part of northern Tien Shan. In: Pouba and Stemprok (Editors) (q. v.), pp. 108-112.

Zonenshain, L. P., 1968. Paleozoic structures of the folded belt of central Asia and their comparison with Middle European paleozoides. In: Mistik (Editor) (q. v.), pp. 305-312.

Zonenshain, L. P., 1971. Geosynclinal process and the new global tectonics. *Geotectonics*, *6*, 3-26.

Zonenshain, L. P., 1972. Similarities in the evolution of geosynclines of different types. *Internat. Geol. Cong. Abstr.*, *24*, 106.

Zonenshain, L. P., 1973. The evolution of central Asiatic geosynclines through sea-floor spreading. *Tectonophysics*, *19*, 213-232.

Zonenshain, L. P., Kuzmin, M. I., Kovalenko, V. I., and Zaltykovsky, A. J., 1974. Mesozoic structural-magmatic pattern and metallogeny of the western part of the Pacific belt. *Earth Planet. Sci. Ltr.*, *22*, 96-109.

Zwart, H. J., 1975. Proterozoic metamorphism in Europe. Mtg. Roy. Soc. Lond., Mar. 13-14.

ABBREVIATIONS USED IN THE ABOVE REFERENCES

Alabama Acad. Sci. J. = Alabama Academy of Science Journal

Amer. Assoc. Petrol. Geol. Bull. = American Association of Petroleum Geologists Bulletin

Amer. Inst. Min. Metal. Petr. Engrs. = American Institute

of Mining, Metallurgy, and Petroleum Engineers

Amer. J. Sci. = American Journal of Science

Amer. Sci. = American Scientist

Ann. Rev. Fluid Mech. = Annual Review of Fluid Mechanics

Austral. Min. = Australian Mining

Ber. Natur. Ges. Freiburg = Berichten Naturforschungen Gesellschaft, Freiburg.

Bol. Comis. Mapa Geol. Esp. = Boletin de la Comision de Mapa y Geologia de Espana

Bull. Geol. Soc. France = Bulletin of the Geological Society of France

Bull. Inst. Roy. Col. Belg. = Bulletin of the Institute of the Royal College, Belgium

Bull. Seis. Soc. Amer. = Bulletin of the Seismological Society of America

Bull. Volcanol. = Bulletin of Volcanology

Bu. Sci. Publ. = Bureau of Science Publication

Can. Inst. Min. Metall. Trans. = Canadian Institute of Mining and Metallurgy Transactions

Can. J. Earth Sci. = Canadian Journal of Earth Science

Can. J. Phys. = Canadian Journal of Physics

Can. Min. Metall(urg). Bull. = Canadian Mining and Metallurgical Bulletin

Can. Min. J. = Canadian Mining Journal

Chem. Geol. = Chemical Geology

Contr. Mineral. Petrol. = Contributions to Mineralogy and Petrology

Dokl. Acad. Nauk SSR = Doklady Academia Nauk USSR

Earth Planet. Sci. Ltr. = Earth and Planetary Science Letters

Earth Sci. Rev. = Earth Science Reviews

Econ. Geol. = Economic Geology

15 Ann. Rpt. Res. Inst. Af. Geol. Leeds = 15th Annual Report

of the Research Institute of African Geology, Leeds

Geochem. = Geochemistry

Geochem. Internat. = Geochemistry International

Geochim. Cosmochim. Acta = Geochimica et Cosmochimica Acta

Geol. Assoc. Can. Spl. Ppr. = Geological Association of Canada Special Paper

Geol. Jahrb. = Geologisches Jahrbuch

Geol. Mag. = Geological Magazine

Geol. Rund. = Geologische Rundschau

Geol. Soc. Amer. Abstr. Prog. = Geological Society of America Abstracts with Programs

Geol. Soc. Amer. Bull. = Geological Society of America Bulletin

Geol. Soc. Amer. Mem. = Geological Society of America Memoir

Geol. Soc. Amer. Spl. Ppr. = Geological Society of America Special Paper

Geol. Surv. Borneo Reg. Malaysia Mem. = Geological Survey of the Borneo Region of Malaysia Memoir

Geol. Surv. Can. = Geological Survey of Canada

Geol. Surv. Can. Spl. Ppr. = Geological Survey of Canada Special Paper

Geol. Surv. Dept. Brit. Terr., Borneo Bull. = Geological Survey, Department of British Territory of Borneo Bulletin

Geophys. Res. Ltr. = Geophysical Research Letters

GSA Spl. Ppr. = Geological Society of America Special Paper

Gulf Coast Assoc. Geol. Soc. Trans. = Gulf Coast Association of Geological Societies Transactions

Hung. Geophys. Inst. Geophys. Trans. = Hungarian Geophysical Institute, Geophysical Transactions

Inst. Min. Metall. Trans. = Institute of Mining and Metallurgy Transactions

Internat. Geol. Cong. Abstr. = International Geological

Congress Abstracts

Internat. Symp. Hydrochem. Biochem. = International Symposium of Hydrochemistry and Biochemistry

Japan. Soc. Min. Geol. = Japanese Society of Mining and Geology

J. Astronaut. Sci. = Journal of Astronautical Sciences

J. Geol. = Journal of Geology

J. Geol. Soc. Austral. = Journal of Geological Society of Australia

J. Geol. Soc. Lond. = Journal of the Geological Society of London

J. Geophys. Res. = Journal of Geophysical Research

J. Petrol. = Journal of Petrology

J. Res. U. S. Geol. Surv. = Journal of Research, U. S. Geological Survey

Koninkl. Ned. Meteorol. Inst. Mededel. en Verh. = Koninklijke Nederlande Meteorologie Institut Mededelungen en Verhandlungen

Nat. Acad. Sci. Mem. = National Academy of Science Memoir

Nature Phys. Sci. = Nature, Physical Sciences

Min. Dep. = Mineralium Deposita

Mtg. Roy. Soc. Lond. = Meeting of the Royal Society of London

Nat. Res. Cncl. Philippines Bull. = National Research Council, Philippines, Bulletin.

Oil Gas J. = Oil and Gas Journal

Ore. Dept. Geol. Min. Res. Bull. = Oregon Department of Geology and Mineral Resources Bulletin

Philippine Geol. = Philippine Geology

Philippine J. Sci. = Philippine Journal of Science

Philippine Weather Bur. Sci. Ppr. = Philippine Weather Bureau Scientific Papers

Phil. Trans. Roy. Soc. Lond. = Philosophical Transactions of the Royal Society of London

Phys. Earth Planet. Int. = Physics of the Earth and Planetary

Interiors

Precamb. Res. = Precambrian Research

Proc. Pac. Sci. Cong., Pac. Sci. Assoc. = Proceedings of the Pacific Scientific Congress, Pacific Scientific Association

Proc. Roy. Soc. Lond. = Proceedings of the Royal Society of London

Proc. USSR Acad. Sci. Geol. = Proceedings of USSR Academy of Sciences, Geology

Publ. Dom. Obs. Ottawa = Publications of Dominion Observatory, Ottawa

Quart. J. Geol. Soc. Lond. = Quarterly Journal of the Geological Society of London

Rev. Geophys. = Reviews of Geophysics

Rept. Internal. Geol. Congr. = Report of International Geological Congress

Roy. Soc. Lond. Phil. Trans. = Royal Society of London Philosophical Transactions

Tectonophys. = Tectonophysics

Trans. Amer. Geophys. Union = Transactions of the American Geophysical Union

Trans. Inst. Min. Metall. = Transactions of the Institute of Mining and Metallurgy

Univ. Chile Comunic. Esc. Geol. = Universidad de Chile Comunicaciones de la Escuela de Geologia

Univ. Tok. Fac. Sci. J. = University of Tokyo Faculty of Science Journal

U. S. Geol. Surv. Prof. Ppr. = U. S. Geological Survey Professional Paper

INDEX